AN INTRODUCTORY GUIDE TO INDUSTRIAL FLOW

Other titles in this series:

An Introductory Guide to Flow Measurement – R. C. Baker
An Introductory Guide to Pumps and Pumping Systems – R. K. Turton
An Introductory Guide to Industrial Tribology – J. D. Summers-Smith
An Introductory Guide to Valve Selection – E. Smith and B. E. Vivian

An Introductory Guide to Industrial Flow

ROGER C. BAKER

Series Editor
Roger C. Baker

Mechanical Engineering Publications Limited, London

First published 1996

This publication is copyright under the Berne Convention and the International Copyright Convention. All rights reserved. Apart from any fair dealing for the purpose of private study, research, criticism, or review, as permitted under the Copyright Designs and Patents Act 1988, no part may be reproduced, stored in a retrieval system, or transmitted in any form or by any means, electronic, electrical, chemical, mechanical, photocopying, recording or otherwise, without the prior permission of the copyright owners. *Unlicensed multiple copying of the contents of this publication is illegal.* Inquiries should be addressed to: The Managing Editor, Mechanical Engineering Publications Limited, Northgate Avenue, Bury St Edmunds, Suffolk IP32 6BW, UK.

ISBN 0 85298 983 0

© Roger C. Baker

A CIP catalogue record for this book is available from the British Library.

The data here are provided in good faith, but the author is not able to accept responsibility for the accuracy of any of the information included, or any of the consequences that may arise from the use of the data or designs or constructions based on any of the information supplied or materials described. The inclusion or omission of a particular material in no way implies anything about its performance with respect to other materials.

The publishers are not responsible for any statement made in this publication. Data, discussion, and conclusions developed by the Author are for information only and are not intended for use without independent substantiating investigation on the part of the potential users. Opinions expressed are those of the Author and are not necessarily those of the Institution of Mechanical Engineers or its publishers.

Typeset by Paston Press Ltd, Loddon, Norfolk
Printed in Great Britain by Ipswich Book Co. Ltd., Ipswich, Suffolk.

SERIES EDITOR'S FOREWORD

As an engineer I have often felt the need for introductory guides to aspects of engineering outside my own area of knowledge. MEP welcomed the concept of an introductory series to follow on from my own book on flow measurement. We hope that the series will provide engineers with an easily accessible set of books on common and not-so-common areas of engineering. Each author will bring a different style to his subject, but some valued features of the original volume, such as conciseness and the emphasis of certain sections by shading, have been retained. The initial volumes are biased towards fluids, but we hope to broaden the scope in later volumes.

The series is designed to be suitable for practising engineers and technicians in industry, for design engineers and those responsible for specifying plant, for engineering consultants who may need to set their specialist knowledge within a wider engineering context, and for teachers, researchers and students. Each book will give a clear introductory explanation of the technology to allow the reader to assess commercial literature, to follow up more advanced technical books and to have more confidence in dealing with those who claim an expertise in the subject.

In this fifth volume in the series, I have attempted to cover the background fluid mechanics, and some thermodynamics, which will enable the users of this series to set the various volumes in context. I hope that it will be as widely read and as useful as the first volume appears to have been and as those of my colleagues are proving to be.

I would particularly like to acknowledge the encouragement, enthusiasm and help of MEP's past and present editorial team, especially Judith Entwisle-Baker, Louise Oldham, and Mick Spencer, to develop the series. We all hope that the series will find a welcome with engineers and we shall value reactions and any suggestions for further volumes in the series.

Roger C Baker
October 1995

CONTENTS

Series Editor's Foreword — v
Author's Preface — xi
Acknowledgements — xii
Nomenclature — xiii

Chapter 1 Introduction — 1
  1.1 Examples of fluid mechanics — 1
  1.2 Flow visualization — 3
  1.3 Local velocity measurement — 5
  1.4 Bulk flow measurement — 13
  1.5 Conclusion — 18

Chapter 2 Commonly Measured Fluid Parameters — 19
  2.1 Temperature — 19
  2.2 Pressure — 27
  2.3 Density — 37
  2.4 Viscosity — 43
  2.5 Surface tension — 46
  2.6 Concentration measurement — 46
  2.7 Level — 48

Chapter 3 Basic Ideas of Fluid Mechanics — 53
  3.1 Hydrostatics — 53
  3.2 Flow similarity — 55
  3.3 Pipe flow profiles — 57
  3.4 Continuity — 62
  3.5 The first law of thermodynamics and the energy equation — 63
  3.6 The second law of thermodynamics and reversibility — 68
  3.7 Derivation of Bernoulli's equation — 71

Chapter 4 Flow Losses in Pipes and Ducts — 73
  4.1 Introduction — 73
  4.2 Loss coefficients — 73
  4.3 Loss coefficient for a straight pipe — 74

4.4	Loss coefficient for inlets	75
4.5	Loss coefficient for bends	75
4.6	Loss coefficient for valves	77
4.7	Calculation of losses for a system	77
4.8	Flow conditioning	79

Chapter 5 Flows in the Boundary Layer next to a Duct Wall — 83
- 5.1 Introduction — 83
- 5.2 Development of the boundary layer — 83
- 5.3 Structure of the turbulent boundary layer — 87
- 5.4 Flow in small gaps — 91

Chapter 6 Compressible Flow — 93
- 6.1 Introduction — 93
- 6.2 Flow in a convergent nozzle — 94
- 6.3 Flow through a convergent–divergent nozzle — 95
- 6.4 Equations governing compressible flow — 98
- 6.5 Equations applied to nozzle flows — 101
- 6.6 Examples of calculations of nozzle flows — 101
- 6.7 Shock waves — 106
- 6.8 Example of a convergent–divergent duct calculation with shock wave — 110
- 6.9 Constant area ducts — 113
- 6.10 Constant area duct with friction–Fanno line — 114
- 6.11 Constant area duct with heat transfer–Rayleigh line — 118

Chapter 7 Oscillations and Waves in Fluids and Water Hammer — 121
- 7.1 Introduction — 121
- 7.2 Surface waves — 121
- 7.3 Pulsation problems in pipes — 123
- 7.4 Water hammer — 123
- 7.5 Fluid oscillations — 125
- 7.6 Ultrasound — 127

Chapter 8 Open Channel Flow — 131
- 8.1 Introduction — 131
- 8.2 Weirs and flumes — 132
- 8.3 Hydraulic jump — 133

Chapter 9 Multiphase Flow — **135**
 9.1 Types of flow — 135
 9.2 Flow from an oil well — 135
 9.3 Horizontal two-phase flows — 136
 9.4 Gas in solution and air entrainment — 138
 9.5 Cavitation — 139
 9.6 Other multiphase flows — 139
 9.7 Flow maps — 140
 9.8 Gas entrapment — 141
 9.9 Bubbles, drops, and particles — 142

Chapter 10 Steam — **147**
 10.1 Introduction — 147
 10.2 Definitions — 149
 10.3 Tables and charts — 150
 10.4 Equations of state — 154
 10.5 Solutions to simple flow problems — 158

Chapter 11 Computer Fluid Dynamics: Present Achievements and Future Developments — **165**
 11.1 Introduction — 165
 11.2 Flow in pipes and pipe components — 166
 11.3 Flow through turbomachines — 170
 11.4 Flow in other engineering applications — 175
 11.5 Flow in flowmeters — 175
 11.6 Future developments — 180

Appendices
 A Gas flow tables — 181
 B Steam tables — 191
 C Steam chart — 193
 D Boundary layer equations — 197
 E Derivation of the compressible flow equations — 211

References — **217**

Bibliography — **221**

Index — **225**

AUTHOR'S PREFACE

This book is aimed at the busy practising engineer who wishes to familiarize himself or herself with the essentials of fluid mechanics as quickly and painlessly as possible. Much of the material in this volume stems from short courses for industry which I have given at Cranfield University. It has, therefore, been frequently tried out on industrial engineers who are usually very discerning in their identification of what they need for their work.

I have attempted to present the essential ideas using straightforward mathematics, and I have introduced links with everyday observations, as well as including some simple engineering calculations.

Bernoulli's equation is central to much of this book and it therefore appears in Chapter 2 to illustrate changes in pressure and velocity. In Chapter 3 Bernoulli's equation is derived from the steady flow energy equation. It also provides the basis for the equation of the Pitot tube in Chapter 1 and will be used in many of the other chapters.

The chapters move from an introduction and discussion of basic measurements, through simple explanation of the essential equations of fluid mechanics, and on to losses, boundary layers, and compressible flows, where some of the fascinating effects are explained in a way which is easy to understand. Chapters on oscillations and open channel flow deal briefly with some occasionally encountered and topical subjects. Multiphase flow is given the merest of reviews, but it leads on to the more specific topic of steam flows. Finally, a brief review of the extremely powerful and rapidly progressing subject of computational methods provides a fitting conclusion to the book.

Some topics have been omitted from this book, either because they are beyond its intended scope, for instance aerofoil theory, or because they might form the subjects of future volumes in the series, for example hydraulic power.

I hope that my treatment will be found easily accessible and that it will be as widely read and as useful as the first volume appears to have been and as those of my colleagues are proving to be.

Roger C. Baker

ACKNOWLEDGEMENTS

The subjects covered in this book have greatly benefited from my industrial experience both in contractual work for many industrial firms, and also in the extremely fruitful discussions during short course lectures which I have given and subsequently while socialising at the end of a hard day of short course lectures and practical work. To all those who have provided insight, I record my thanks.

I have drawn on some of the experimental and computational work in which I have collaborated over the years and I acknowledge my debt to many colleagues and in particular to Apostolos Goulas who initiated the course and encouraged me to write the lectures which form the basis of this book.

The Hampshire Technology Centre have kindly provided Fig. 1.1 which was obtained from one of the wind tunnels in their interactive centre for schools. Northern Research and Engineering Corporation have kindly given permission for the use of Fig. 1.2 which was obtained as part of a jointly funded programme on centrifugal compressor performance. Figures taken from the Journal of Flow Measurement and Instrumentation are referenced at the end of the book and are reprinted with kind permission from Butterworth–Heinemann journals, Elsevier Science Limited, The Boulevard, Langford Lane, Kidlington OX5 1GB, UK. Figures taken from the Proceedings of the Institution of Mechanical Engineers are referenced at the end of the book and are reprinted with kind permission. I am also grateful to Cambridge University Engineering Department for permission to include the gas tables and to Cambridge University Press for permission to include the density chart from the steam tables by Hayward. The part of the enthalpy-entropy chart for steam included in this book is reproduced with kind permission from Basil Blackwell, Oxford.

I am most grateful to Don Miller for reading through the manuscript and making many helpful suggestions, many of which I have incorporated. I should stress, however, that he bears no responsibility for errors in the book. Lastly I would like to thank my family, and particularly Liz, my wife, who has put up with my unsociable hours sitting at a word-processor typing this out.

NOMENCLATURE

A	Cross-sectional area; constant; coefficient in Beattie–Bridgman equation
a	Constant in Van der Waal and Beattie–Bridgman equations
B	Constant; coefficient in Beattie–Bridgman equation
b	Distance between centre of gravity and centre of buoyancy; width of channel; constant in Van der Waal and Beattie–Bridgman equations
C	Constant; concentration of water in an oilflow
$\bar{C}$	Mean concentration
C_D	Drag coefficient for drops, bubbles, and particles
c	Speed of sound; constant in Beattie–Bridgman equation
c_p	Specific heat at constant pressure
c_v	Specific heat at constant volume
D	Pipe diameter
F	Impulse function–remains constant in Rayleigh line flows
Fr	Froude number
f	Friction coefficient
f_D	Friction coefficient used in Fig. 4.2 and equation (4.6) ($f_D = 4f$)
f_{circ}	Factor for circulation in droplet
g	Gravitational acceleration (9.81 ms^{-2})
H	Height of still free liquid surface above datum
h	Height of manometer column; specific enthalpy; head over the weir
I_o	Emergent intensity of radiation
I_i	Incident intensity of radiation
K	Constant; loss coefficient
k	Coefficient for pitot tube; spring constant; roughness
L	Length of resistive element; length of pipe over which pressure loss occurs; length of small gap
L_{MAX}	Maximum length of duct with friction to cause sonic conditions at exit
l	Length of tube of fluid in laminar flow profile calculation; distance from the leading edge of the plate in boundary layer estimations

M	Mass of fluid container; righting moment; Mach number
$\mathbf{M}$	Molecular weight (29 for air)
m	Mass of fluid in a closed system
n	Shear strain index; turbulent flow profile index
P	Power transmission (ultrasound)
p	Pressure; static pressure
p_a	Ambient pressure
p_b	Back pressure
p_c	Pressure at the critical point
p_{crit}	Critical pressure in compressible flow
p_e	Exit pressure
p_g	Gauge pressure
p_o	Stagnation or total pressure
p_R	Relative pressure
p_t	Throat pressure
Q	Heat transfer
q_v	Volumetric flow rate
q_m	Mass flow rate
R	Gas constant for a particular gas (286.7 J kg^{-1} K^{-1} for air); electrical resistance; tube radius
$\mathbf{R}$	Universal gas constant related to molecular weight in kilograms (8.3143 kJ kmol^{-1} K^{-1})
Re	Reynolds number
r	Radial position; radius of inlet lip; extensive property of steam
r_f	Saturated liquid value for extensive property of steam
r_g	Saturated vapour value for extensive property of steam
S	Entropy
s	Specific entropy (per unit mass)
T	Temperature
T_c	Temperature at the critical point
T_o	Stagnation or total temperature
T_R	Relative temperature
t	Width of small gap
u	Internal energy
V	Velocity
V_o	Centre line velocity
$\bar{V}$	Mean velocity

Nomenclature   xv

V_∞	Free stream velocity
v	Specific volume; wave velocity
v_o	Specific volume at 0°C
W	Work transfer; terminal velocity for drops, bubbles and particles
X	Variable used in the formula for temperature
x	Movement in the thermometer stem; movement of fluid container; distance traversed by radiation; mass fraction of vapour in steam
Y	Variable used in the formula for temperature
Z	Impedance (ultrasound)
z	Height above datum; valve opening

Greek symbols

α	Temperature coefficient for platinum resistance thermometer and thermistor; constants in steam equation
β	Thermal expansion coefficient (sometimes α is used)
γ	Ratio of specific heats (c_p/c_v)
Δp	Pressure difference
$\Delta \rho$	Difference between the density of water and oil
δ	Boundary layer thickness
δm	Small element of mass
ϵ	Constant in Beattie–Bridgman equation; eddy viscosity
η	Efficiency
κ	Ratio of viscosity for droplet μ_p/μ; bulk modulus
λ	Damping coefficient; wavelength
μ	Mass absorption coefficient; dynamic viscosity; equal to the ratio pv/RT
ν	Kinematic viscosity
ρ	Density; coefficient of electrical resistivity
ρ_o	Stagnation density; coefficient of electrical resistivity at T_0
ρ_m	Density for mercury
ρ_w	Density for water
σ	Surface tension
τ	Shear stress
ω_r	Resonant frequency

Subscripts

a	Ambient
b	Back
c	Critical
CIRC	Circulation
D	Drag; Darcy
e	Exit
f	Value for saturated liquid
fg	Difference between a saturated liquid value and a saturated vapour value as in $h_{fg} = h_g - h_f$
g	Gauge; value for saturated vapour
i	Inlet
m	Mercury; mass (flow rate)
MAX	Maximum
p	(Specific heat at) constant pressure
R	Relative; reflected
r	Resonant
s	Shock; constant entropy
T	Constant temperature; transmitted
t	Throat
v	(Specific heat at) constant volume; volumetric (flow rate)
w	Water
z	Axial component
θ	Angular component
0	Stagnation; centre line
1	Station 1; medium 1
2	Station 2; medium 2
3	Station 3
∞	Free stream

to Liz, Sarah and Paul, Mark, John and Rachel

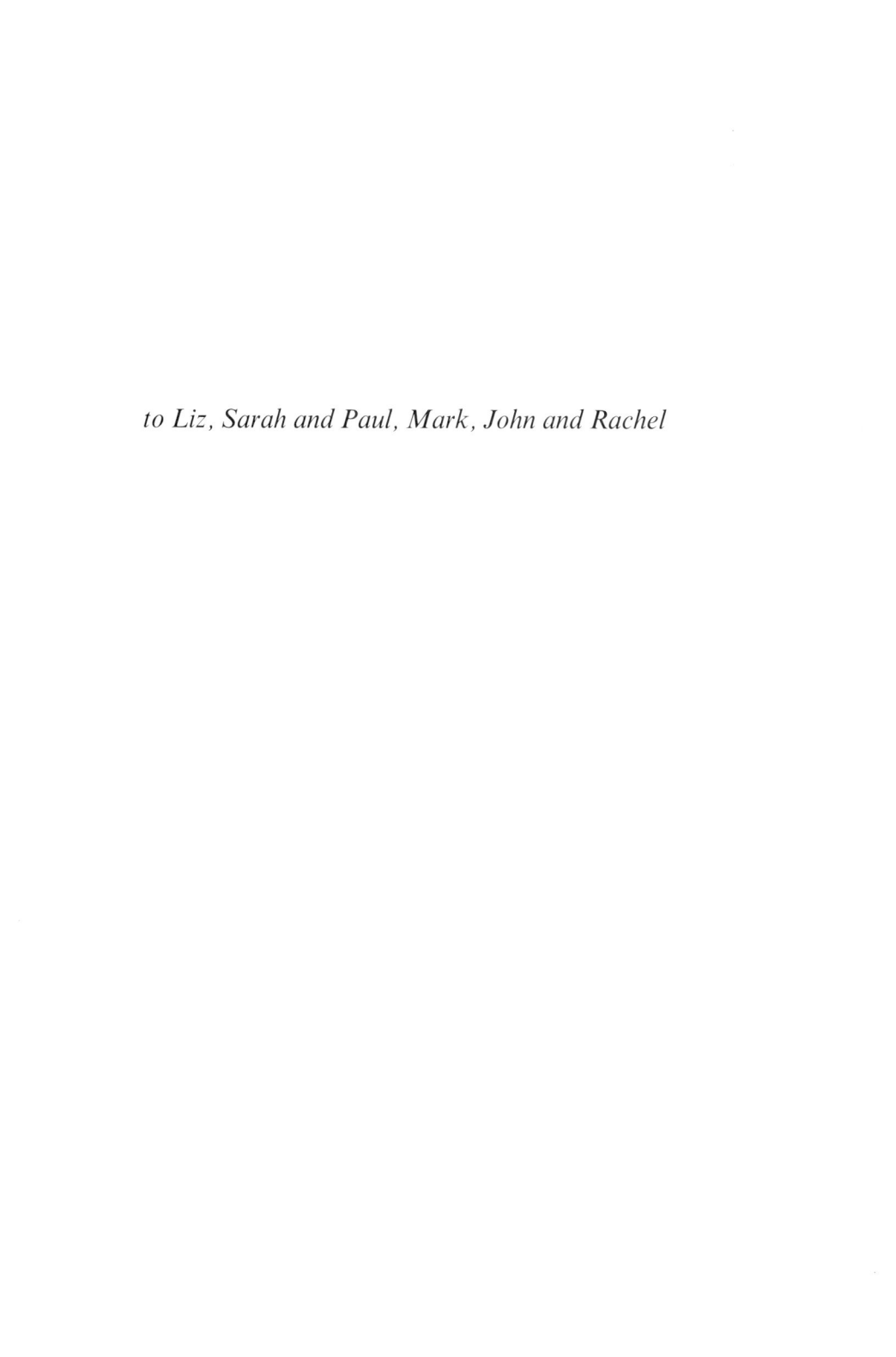

CHAPTER 1

Introduction

> The all-pervading nature of fluid mechanics
> The need to understand its effects
> The ways of visualizing the flow
> The ways of measuring the local velocity of fluid
> A brief review of bulk flow measurement

1.1 EXAMPLES OF FLUID MECHANICS

A thorough grasp of fluid mechanics and thermodynamics is not only essential to an understanding of flow in industrial engineering systems, flow measurement, pumps, pipes and valves, around ships and aeroplanes and – on a more domestic level – in vacuum cleaners, heaters, and air conditioners, but it also gives a fascinating insight into many phenomena in the world around us. Most people will appreciate the development which has gone into creating our modern aircraft. The work involved in producing efficient delivery systems for water, gas, and oil may be less exciting, but is just as important for the efficient use of resources and the protection of the environment. A better understanding of the fluid engineering of a vacuum cleaner or a washing machine might lead to savings in energy and time, and to reduced levels of noise in the home. We take for granted the weather systems which create lows or depressions or highs and anticyclones and which can result in violent storms; it is forgotten that these systems are due to the movement of air from high to low pressure regions on a planet which is rotating. The same effect is used in coriolis flowmeters (**1**). The vortex shedding which is a common feature of flow past a bluff body is to be found around pipes and cables used in North Sea oil operations, in the flow around tall chimneys, and in the operation of the vortex shedding meter. In the first two examples we need to know how to control the phenomena which may cause disaster, while in the third we seek to enhance the same phenomenon to improve the signal size. And why can we hear sound better downwind than upwind? Lamb (2) pointed out that this is not due to the sound being carried downwind better than upwind, but because of the focusing effect of the boundary layer on the earth's surface. However, in the ultrasonic flowmeter the sound is certainly carried by the flow and the time difference, although not the strength of the source, is a key feature in the measurement (1).

In this book we shall use the term fluid to mean liquid or gas, and will only refer to one or the other when the more general term does not apply.

The adjective "streamlined" is easily used, but what is meant by it? Adverts for new sporty cars are frequently seen with lines around them, suggesting the flow of the air passing smoothly over them, but does it really and, if so, how do we know? If a streamlined shape is one where the flow follows the contour of the shape, what determines this? Often flow can be observed in everyday surroundings, some examples of which are listed below.

- The beautiful cloud patterns are evidence of thermal convective flows combined with winds, so that as the humid air rises it causes condensation.
- When following another vehicle in wet weather one can sometimes observe the twin vortices behind it.
- The trailing vortices off the wing tips of a jet aircraft will become visible for a few seconds as they cause local precipitation of water particles.
- On a fine night, with enough phosphorescence in the sea water, the bow wave from a ship or from flow around a submerged object can be seen.
- The walls of subway tunnels show the regions where recirculation has deposited dirt particles on the wall.
- Drifting snow can create the most fantastic shapes as it follows the recirculation of the blizzard downstream of an obstruction.
- Film sequences of spectacular explosions and fires look as if they are modelled at small-scale rather than full size. Why is this? Our discussion of Reynolds number may throw some light on this question.

It would be nice to cover all these phenomena, and more, in this book, but space will require that the treatment is selective.

1.2 FLOW VISUALIZATION

As a starting point it is useful to look at how the flow pattern around any shape can be artificially visualized. This is a first step for the engineer when trying to understand what is happening in any particular application. There are many means of flow visualization, and the books in the flow visualization section of the bibliography provide some brilliant examples.

Some of the more common means of flow visualization are:

For air:	Smoke and other particles
	Tufts
	Surface paint
Air with temperature and density gradients:	Optical methods such as schlieren, interferometry and holographic techniques.
Water:	Hydrogen bubbles
	Dye
	Tufts
	Fluorescence
	Aluminium particles on the surface
Wear due to solids in water:	Paint layers of various colours

Figure 1.1 illustrates the use of smoke to show the flow patterns. Figures 1.1 (a) and (b) illustrate, on the one hand, the benefits of a reasonably good streamline shape for a car, and on the other, the disturbance from a roof rack and consequent increased drag.

Figures 1.1(c) illustrates the use of smoke to show flow patterns over a house. The separation of the flow after the house ridge contributes to a low pressure region on the downwind roof of the house and a resulting upward force on the roof. The reason for this is, as will be seen with the venturi tube, that as the velocity of the fluid increases, the pressure in the stream falls. The same pressure will exist below the fast stream line in the recirculation region, hence causing the low pressure.

(a)

(b)

Fig. 1.1 Smoke visualization of flow (a) around a car; (b) around a car with a loaded roof rack.
(*By courtesy of Hampshire Technology Centre*)

Introduction

(c)

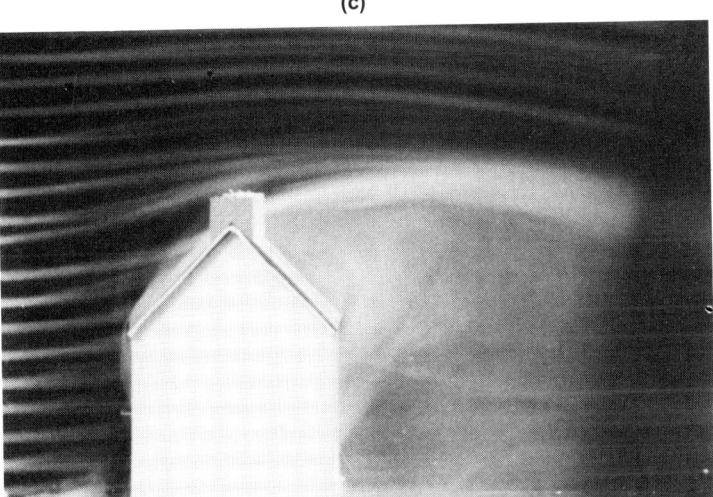

Fig. 1.1 Smoke visualization of flow (c) over a model house.
(By courtesy of Hampshire Technology Centre)

Figure 1.2 shows the use of hydrogen bubbles to visualize the flow around a centrifugal impeller. The fine arrays of platinum wires as cathodes can be seen in Fig. 1.2(a) and the sheets of bubbles in Fig. 1.2(b). These are taken from some of the author's early work and show the problem of achieving a clear trace in a complex flow such as this one. Tufts as shown in Fig. 1.2(a) at the impeller outlet can sometimes show local flow directions. Figure 1.3 shows the wear patterns obtained in a slurry-carrying pump by using paint layers which were worn away by the slurry particles in the flow. This piece of research will be returned to in Chapter 11 where it will be shown how computation can help to predict this erosion.

1.3 LOCAL VELOCITY MEASUREMENT

Visualization methods are increasingly becoming more than qualitative and are being used to obtain quantitative measurements of velocity across a flow field. However, there are three common methods for obtaining local velocity which have been used by experimental fluid dynamicists.

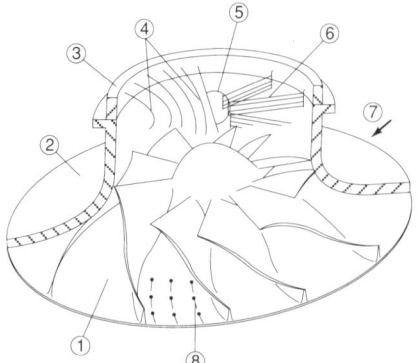

1 Impeller
2 Plexiglass shroud
3 Electrode ring
4 Anode wires
5 Cathode support wire
6 0.0015 inch cathode wire
7 Direction of photographs
8 Tufts (attached at upstream end)

(a) Diagram of the impeller housing to show details of the hydrogen bubble wires and the tufts

(b) Hydrogen bubble sheets

Fig. 1.2 Flow around a centrifugal impeller: visualization using hydrogen bubbles in water flow (3)
(*Courtesy of Northern Research and Engineering Corporation, Ma, USA*)

Common methods of local velocity measurement:
- the pitot tube;
- the hot wire anemometer;
- the laser doppler anemometer (LDA).

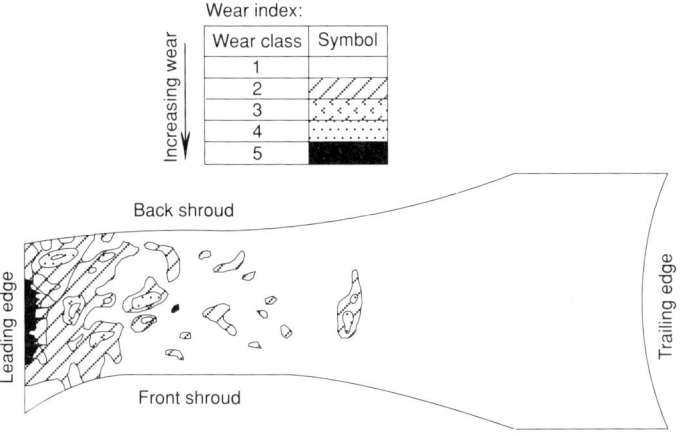

Erosion pattern on blade pressure surface number 3 for 11.7 percent concentration

Fig. 1.3 Wear patterns on the blades of a pump impeller due to slurry particles in the flow shown by erosion of the paint layers (4)

The pitot tube was first suggested by Henri Pitot in 1732 and depends on the fact that when the tube (Fig. 1.4(a)) faces into the flow, the flow comes to rest in the hole and in doing so causes a rise in pressure to the stagnation or total pressure (for which $k = 1$):

$$\Delta p = k \frac{\rho V^2}{2} \qquad (1.1)$$

where ρ is the fluid density and V is the fluid velocity. If the original pressure in the stream, the static pressure, is known, then the difference between this and the stagnation pressure gives Δp and this can be used to calculate the value of V provided ρ is known. The derivation of this equation will be discussed in Chapter 3 (equation (3.29)). One method of obtaining the static pressure is to measure the pressure in the wall of the duct through which the flow is passing, as in Fig. 1.4(a). A better method of obtaining the static pressure is to use a pitot-static tube (Fig. 1.4(b)) in which the static pressure is obtained from pressure tappings on the side of the tube and the two pressures are led down the inner and outer concentric ducts which make up the pitot-static tube.

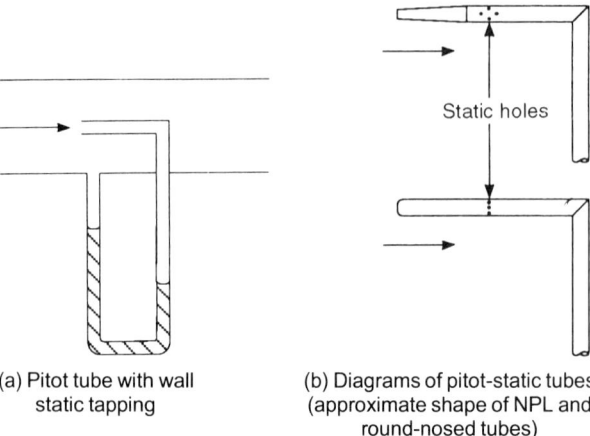

(a) Pitot tube with wall static tapping

(b) Diagrams of pitot-static tubes (approximate shape of NPL and round-nosed tubes)

Fig. 1.4   Pitot tubes

For many years the pitot-static tube was the mainstay of a fluid dynamics laboratory, giving a reading in steady flow very close to the theoretical value in equation (1.1) with $k = 1$. The author made use of a pitot probe (**5**) to obtain the boundary layer shape (see Chapter 5) next to a flat plate in the region where the stationary plate causes shearing in the flow and a decrease in velocity as the plate is approached, Fig. 1.5. The

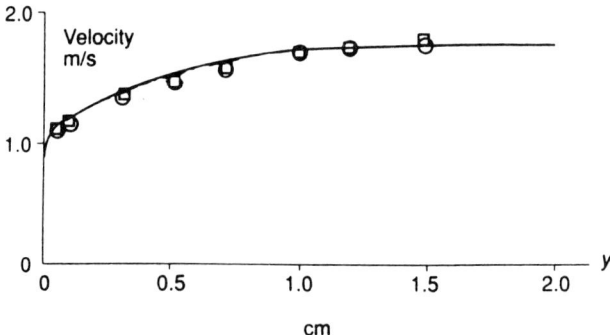

Fig. 1.5   Boundary layer on a flat plate at various stations obtained with a pitot tube (5)

Introduction

boundary layer was measured on a flat plate in a water channel. The author has also used a pitot comb (Fig. 1.6) to measure the velocity simultaneously at a number of points in a boundary layer on a flat-bottomed boat (**6**). Chapter 3 explains the Reynolds number (Re) which is a dimensionless parameter relating the importance of fluid inertia to fluid viscosity. At low values of Re (<100), when the effect of viscosity becomes important, k diverges from unity. In a shear flow there is likely to be a shift in the effective measurement point. Goldstein (**7**) gives k values for National Physical Laboratory (NPL) and round-nosed designs which show a k value of 1.020 (NPL) and 1.055 (round-nosed) at about 0.6 m/s ($Re \approx 300$), a minimum value of $k = 0.991$ at about 3–3.5 m/s (Re 1600–1900) and a value within about 0.1 percent of unity for 6 m/s–27 m/s (Re 3000–10 000). Otherwise the uncertainty can be obtained from BS 1041 Part 2.1.

The most important disadvantage of the pitot tube is the effect which unsteady flow has on its reading, and since turbulence (about which more later), is very common in normal flows, the pitot tube will tend to read high (**7**).

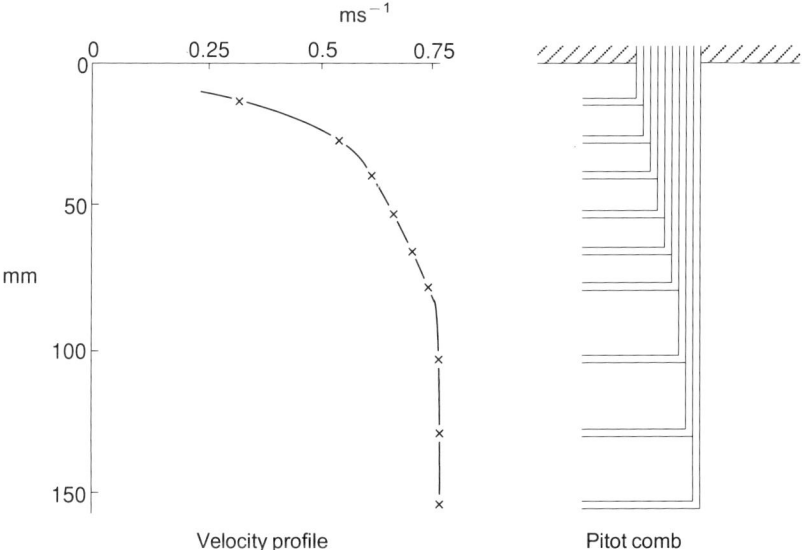

Fig. 1.6 Boundary layer on a flat bottomed boat as measured by a pitot comb (6)

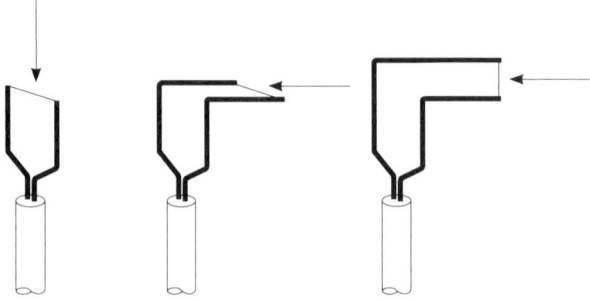

Fig. 1.7   Hot wire probes

For the reasons given above, the hot wire anemometer has been an essential addition to the instrumentation for flow mapping. It operates on the principle that as air (the usual medium in which it is used) flows past a heated wire, the wire is cooled, and the added heat input or the degree of cooling provides a means of deducing the velocity of the passing stream. The wires used in these devices are extremely fine and therefore very fragile. Typical designs are shown in Fig. 1.7. Some experimental results,

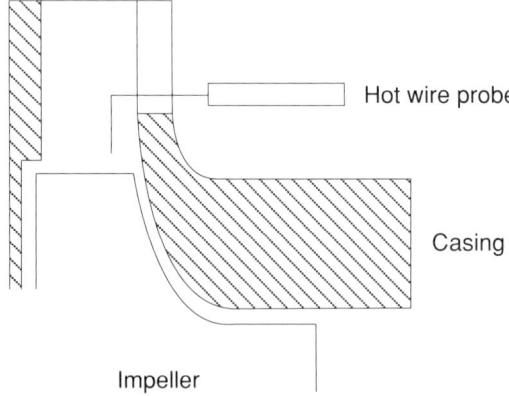

(a) Meridional view of the setting of the hot wire

Fig. 1.8   Measurements at the exit of a centrifugal compressor impeller using hot wire probes (8)

Introduction 11

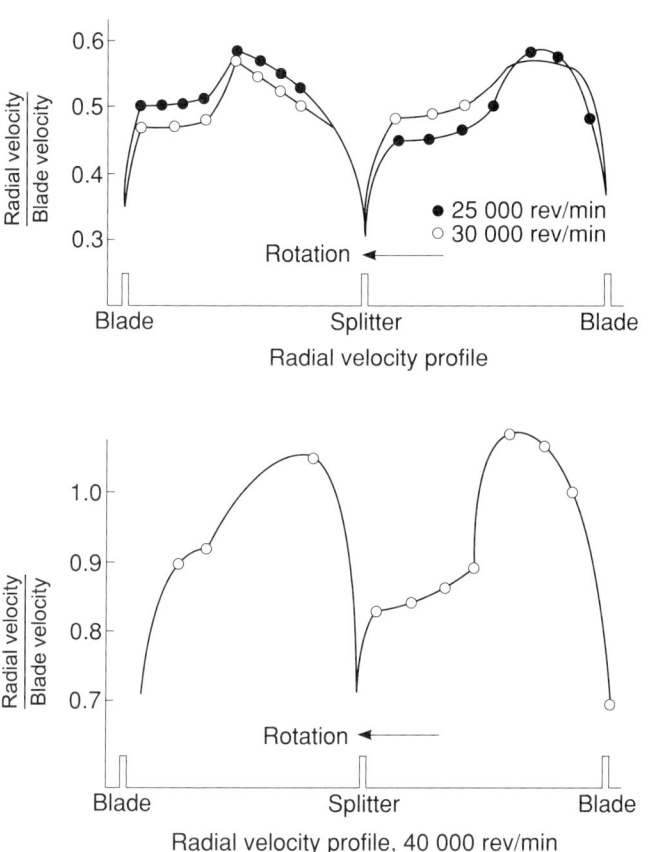

(b) Radial velocity measurements

Fig. 1.8 Measurements at the exit of a centrifugal compressor impeller using hot wire probes (8)

taken by a colleague of the author at the exit of a centrifugal compressor impeller, are shown in Fig. 1.8 (**8**).

Another design of hot wire anemometer makes use of two spaced-out wires, and measures the transit time from one to the other by sending a heated pulse of air from the upstream wire.

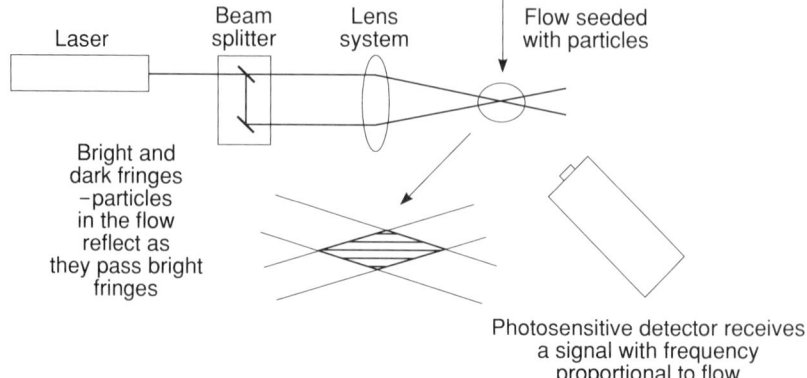

Fig. 1.9 Laser doppler anemometer

The laser doppler anemometer (LDA) offers an elegant non-intrusive method of measuring local velocity in air and water where there is sufficient seeding material to reflect the light. There are two different methods of operation for the LDA. The most common operates by splitting the beam of light and then crossing the beams to create an interference pattern of fringes (Fig. 1.9). As the particles in the flow pass through the pattern they reflect light at a frequency related to the fringe spacing and the velocity of the particles. The other type of LDA operation uses two spots and measures the time of transit of particles between the spots. This is similar to the first type in the sense that the fringe pattern has been replaced by two spots.

Some of the strengths and limitations of this tool are shown in Fig. 1.10. This Figure shows the results of measurements in high velocity air flow with seeding which was not capable of following the rapid acceleration and deceleration of the flow through a shock wave. Another example of the use of this device and its ability to measure complex flows is illustrated in Fig. 1.11 which shows the flow patterns in the region downstream of a 'T' junction where mixing of the flow takes place.

Introduction

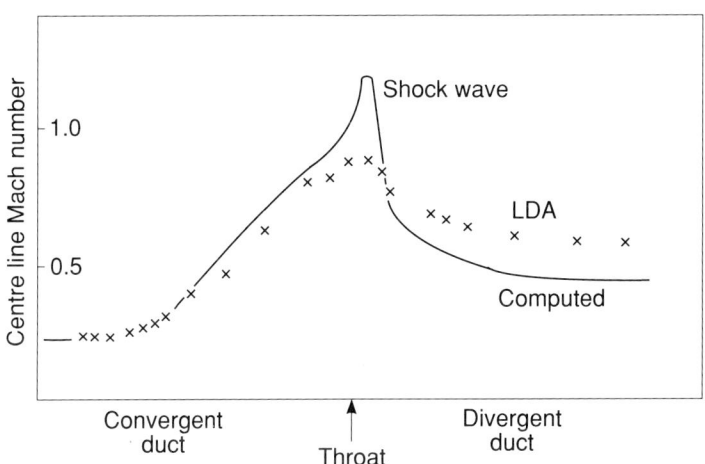

Fig. 1.10 High speed flow with shock wave
(*Courtesy of J.A. Damia Torres*) (9)

1.4 BULK FLOW MEASUREMENT

Momentum:	Orifice, venturi, variable area
Volumetric:	Positive displacement, turbine, vortex, electromagnetic, ultrasonic
Mass:	Thermal, coriolis

Only a brief review of bulk flow measurement is given here, since another volume in this series is dedicated to the topic (**1**), and this review is included merely for completeness. Instruments may broadly be divided into momentum-sensing (for which the density of the fluid is needed to deduce volumetric or mass flow rate), volumetric flow sensing, and mass flow sensing. Figure 1.12 gives a simple diagram of some common types of flowmeter, without attempting to illustrate the many variations and less used designs.

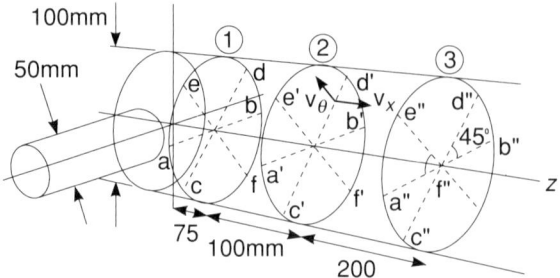

(a) Diameters along which velocity components were measured

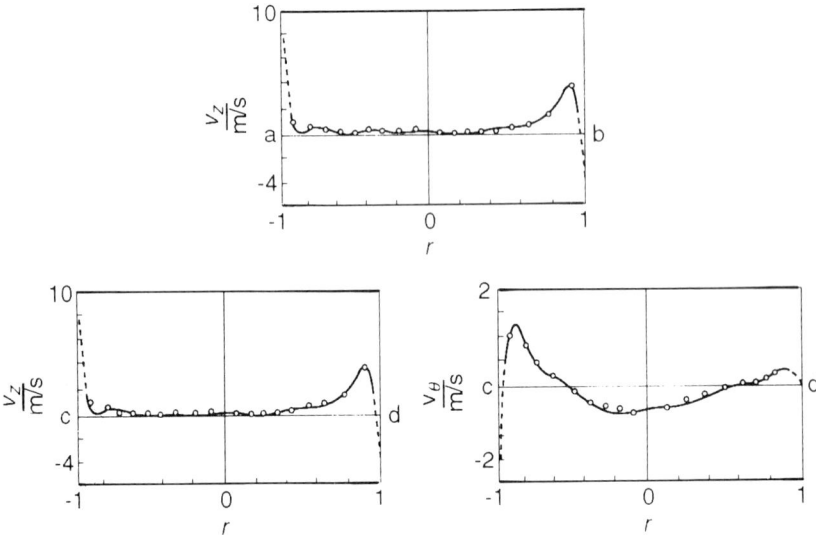

(b) Axial and tangential velocity profiles at station 1 against non-dimensional radius

Fig. 1.11 Flow in a 'T'-junction (10)

Introduction

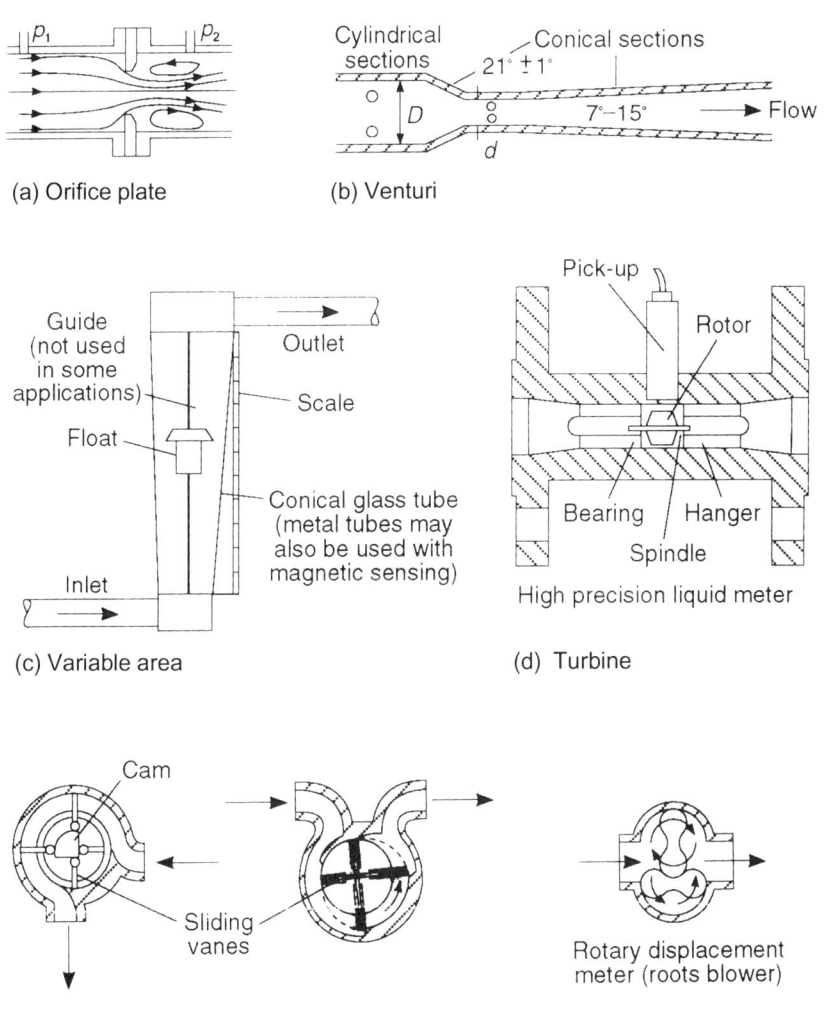

Fig. 1.12 Examples of some common types of bulk flowmeter (1)

16   *An Introductory Guide to Industrial Flow*

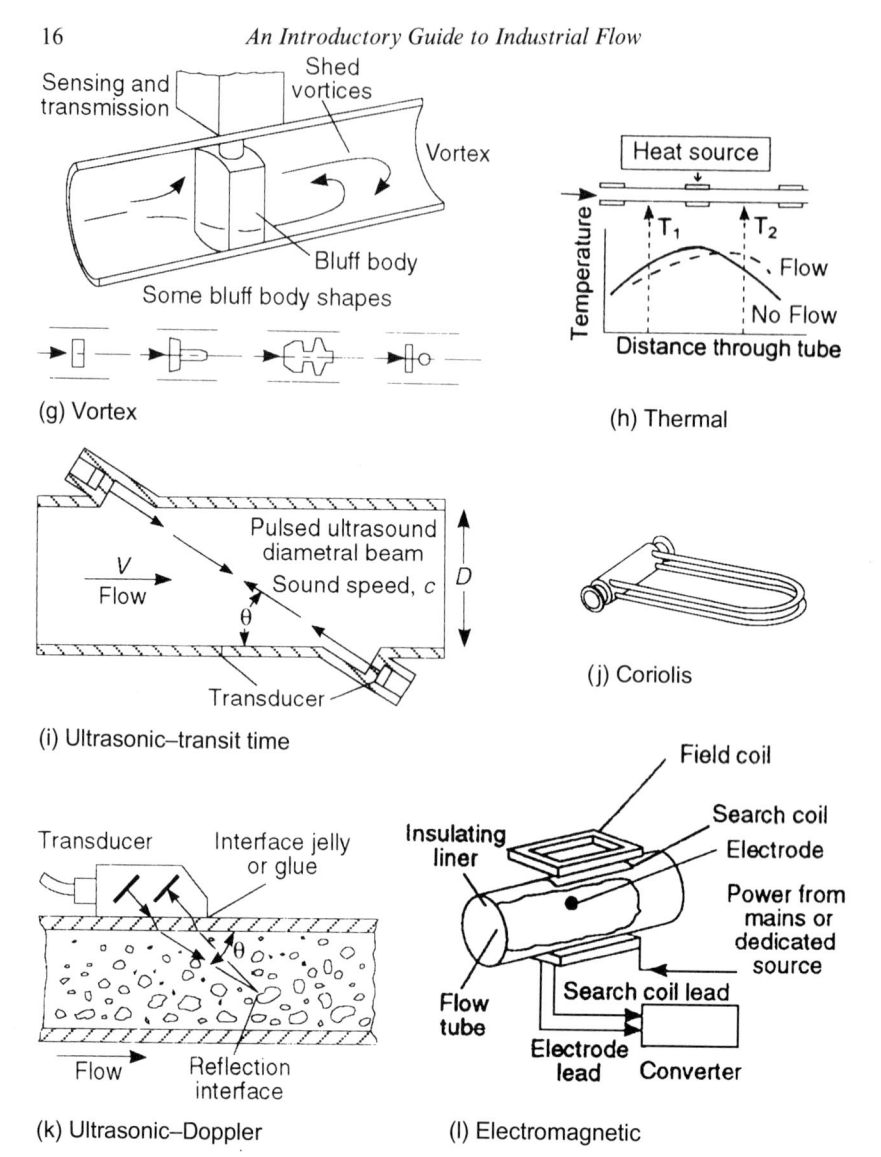

(g) Vortex

(h) Thermal

(i) Ultrasonic–transit time

(j) Coriolis

(k) Ultrasonic–Doppler

(l) Electromagnetic

Fig. 1.12 *Continued*

Momentum

The orifice plate, Fig. 1.12(a), is usually held between pipe flanges. It is of thickness ⩽0.05 of the pipe diameter with a hole of diameter usually less than 80 percent of the pipe diameter. Flow of liquid or gas is constricted and accelerated through the orifice so that the pressure in the flow drops and the change in pressure across the orifice is used to deduce the flow rate. The precision is of order 1 percent, based on the standard design.

The venturi, Fig. 1.12(b), has a smooth contraction and the subsequent pressure recovery is good so that it is used where little pressure loss in the pipe can be accepted for either liquids or gases. The pressure difference for deducing the flow is between inlet and throat. The precision is of order 1 percent, based on the standard design.

The variable area flowmeter, Fig. 1.12(c), has a float which is lifted up in a conical tube by the upward flow of liquid or gas. As the flow increases the float rises higher in the tube, so giving a larger annulus around the float for the fluid to pass through. The height of the float gives the flow rate. Precision is of order 2 percent and is highly fluid-dependent.

Volumetric

In the positive displacement flowmeter the fluid is divided into compartments by the rotors and is carried through the meter. The flow rate is deduced from the revolutions of the rotors. Figure 1.12 shows versions for liquid (e) (precision of order 0.1 percent) and gas (f) (precision of order 0.5 percent).

In the turbine flowmeter a propeller in the flow rotates as the fluid passes and the speed of rotation allows the volumetric flow rate to be deduced. Figure 1.12(d) shows a typical liquid flowmeter (precision of order 0.2 percent), although high-precision (of order 0.5 percent) gas meters are also available.

A bluff body across the pipe of the vortex flowmeter, Fig. 1.12(g), sheds vortices at a frequency proportional to the flow rate of the fluid (equation (7.6)). Precision is of order 0.5 percent.

The electromagnetic flowmeter, Fig. 1.12(l), depends on the observation that, when a fluid flows through a magnetic field, a voltage is generated. In the electromagnetic flowmeter the fluid is almost invariably a conducting liquid and the tube has an internal insulating liner so as not to short out the voltages. Electrodes on each side of the tube sense the voltage which is then amplified. Precision is of order 0.5 percent.

There are two main types of ultrasonic flowmeter.

(a) The doppler meter, Fig. 1.12(k), depends on reflection of the ultrasound off moving particles in the flow, creating a doppler shift in the frequency of the ultrasound which is used to obtain a measure of velocity. Precision is poor.
(b) The transit time meter, Fig. 1.12(i), depends on measuring the time taken for pulses of ultrasound to cross the pipe at an angle to the flow. Gas and liquid versions are available. Precision is of order 1 percent.

Mass

Thermal flowmeters, Fig. 1.12(h), vary widely in design, but all depend on the temperature change in a· flowing fluid, (caused by heat addition) to provide a measure of the mass flow rate. In some devices a heated probe is kept at constant temperature and the heat flow is measured. Liquid and gas versions are available. Precision is of order 1 percent.

The coriolis flowmeter, Fig. 1.12(j), depends on a little known acceleration to sense mass flow. A tube is vibrated so that for part of its length it is experiencing a rotational motion. This, combined with the flow of the fluid along the rotating tube, results in a coriolis acceleration which in turn creates a force that can be measured to give the mass flow rate. These meters are used for liquids and gases and precision as high as 0.2 percent is claimed for them.

To measure granular solids flow, use has been made of impact of flow on a plate and of electrostatic sensing methods.

Open channel flowmeters are briefly dealt with in Chapter 8.

1.5 CONCLUSION

Some of the common examples of fluid mechanics have been reviewed, some of the means of visualizing the flows and measuring them have been described, and bulk flow measurement has been briefly reviewed. The measurements which can be made in flows in addition to these are pressure and temperature. Other properties of the fluid can be measured, such as viscosity, density, and surface tension. The next chapter looks at the instrumentation needed to make these measurements.

CHAPTER 2
Commonly Measured Fluid Parameters

In this chapter the main fluid parameters and methods of measurement are considered

- Temperature
- Pressure
- Density
- Viscosity
- Surface tension
- Concentration
- Level

For more detailed information on these measurements the reader is referred to handbooks such as that by Noltingk (**11**). The following sections give the reader an introduction to the subject.

2.1 TEMPERATURE

In this section the following are considered

- Temperature scales
- Types of thermometer and limits of use
    - expansion of solids
    - liquid in glass
    - liquid in metal
    - gas thermometer
    - platinum resistance
    - thermistor
    - thermocouple
    - other types

Temperature scales

The SI unit of temperature, T, is the degree Celsius (°C) or on an absolute scale the Kelvin (K). It is compared below with the Fahrenheit scale (°F).

Table 2.1 Temperature scales

	°C	K	°F
Absolute zero	−273.15	0	−459.67
Ice point	0	273.15	32
Ambient range	10	283.15	50
	20	293.15	68
	30	303.15	86
	40	313.15	104
Boiling point	100	373.15	212

Thus to move from X degrees Celsius to Y degrees Fahrenheit requires the expression

$$Y = 1.8 \times X + 32 \tag{2.1}$$

and vice versa

$$X = \frac{Y - 32}{1.8} \tag{2.2}$$

It is important to remember that errors in temperature measurement need to be given in °C (or °F) or as a percentage of temperature difference.

Types of thermometer and limits of use

The common methods of temperature measurement which are in use outside the laboratory are listed in Table 2.2 (see reference (**12**) for a fuller discussion).

Expansion of solids

These instruments depend on the expansion of a metal such as brass and often require a second metal of low thermal expansion such as invar. The bimetallic strip (Fig. 2.1) is a common example in which invar and brass are bonded together and expansion of the brass causes the strip to curl in the direction of the invar strip. A common use of this is to provide a thermal cut-out for appliances such as toasters and tumble driers.

Commonly Measured Fluid Parameters

Table 2.2 Thermometer ranges

Types of thermometer	Approximate range °C
Expansion of solid	to +300
Mercury in glass	−40 to +650
Other liquids	−80 to +300
Gas	−130 to +540
Vapour	−50 to +320
Platinum resistance	−250 to +700
Thermistor	−100 to +300
Thermocouples	−200 to +1700

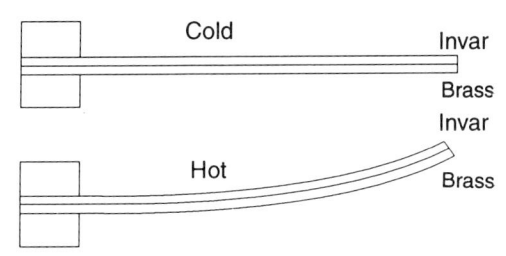

Fig. 2.1 Expansion of solids – bimetallic strip

Liquid in glass

Mercury in glass is the most common combination. This consists of a glass tube with a small capillary up it. The tube may be shaped so that the capillary is magnified by the glass around it for easier reading. At one end is a reservoir to hold an amount of mercury which is large compared to that in the capillary. The other end is sealed with a vacuum above the mercury. If we use the thermal expansion coefficient β (α is sometimes used) in the form

$$\beta = \frac{1}{v}\frac{dv}{dT} \tag{2.3}$$

where v is the volume per unit mass or specific volume, and rewrite it for a temperature change from, for example, 0°C to T°C, then

$$\frac{v - v_o}{v_o} = \beta(T - 0) \tag{2.4}$$

If $v - v_o$ is small compared with v_o, or in terms of the volumes in the thermometer Adx (A is the cross-section of the thermometer tube) is small compared with the volume of the reservoir, then the distance x up the stem will be approximately proportional to the temperature difference.

The precision of the thermometer depends on the precision of the capillary, the precision with which the graduations have been scribed, and the precision of the person who reads the stem. It should also be remembered that glass is not absolutely stable and unchanging, and that with use and age the thermometer may change in calibration. The thermometer should, therefore, be regularly recalibrated, as should every other instrument, if high precision is to be maintained. In use it is important to immerse as much of the thermometer as for the calibration, to ensure correct temperature distribution in the mercury and the glass. This may be more important at higher temperatures where there is a greater heat loss from the stem. Hagart-Alexander has written a very useful article on these thermometers (**12**) and suggests 0.1K as the uncertainty for general purpose instruments.

Other liquids used in glass thermometers are alcohol, toluene, pentane, and creosote (**12**). In cases where mercury is not suitable, for instance where breakage could result in contamination, alcohol and other liquids are used.

Liquid in metal

There are also other forms of mercury-filled thermometers such as Bourdon tubes (see Fig. 2.7a) in which, due to the increase in liquid volume, the flattened tube becomes elliptical, uncurls, and so moves a pointer, as in the equivalent pressure device described in more detail below. Other forms of container are used for mercury. It should not be expected that these instruments will necessarily give particularly high precision.

Gas thermometer

The gas thermometer depends on the equation for an ideal gas (Chapter 3)

$$pv = RT \qquad (2.5)$$

where p is the pressure, v is the specific volume (volume per unit mass), R is the gas constant for a particular gas and T is the temperature. Provided the gas approximates to an ideal gas, so called because the

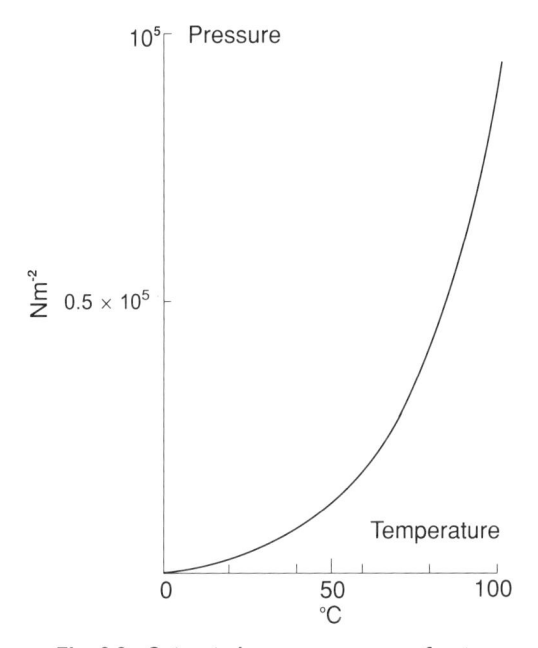

Fig. 2.2 Saturated vapour pressure of water

value of R is constant, and provided the volume of the vessel which constitutes the thermometer is constant, the pressure will be proportional to the temperature. The Bourdon tube is commercially used as a vessel with approximately constant volume.

Another form of this instrument is the vapour pressure thermometer which depends on the fact that wet vapour has a simple and unique functional relationship between pressure and temperature irrespective of volume. Thus if we measure the pressure, say by using a Bourdon tube, a measure of the temperature can be obtained. Figure 2.2 shows a plot of the relationship, which can be seen to be non-linear.

Platinum resistance
Although nickel and copper are also used, platinum has become an industry standard.

Platinum has three particular advantages:

(a) it is very stable;
(b) it has a high coefficient of resistivity;
(c) it has an average-to-high temperature coefficient of resistivity.

The resistance is given by

$$R = \rho L/A \qquad (2.6)$$

where L is the length of the resistive element, A is the cross-sectional area, and ρ, the coefficient of resistivity, may be written in terms of temperature as

$$\rho = \rho_o\{1 + \alpha(T - T_o)\} \qquad (2.7)$$

where α is the temperature coefficient ($\simeq 0.0039/°C$ in the range 0–100°C) and ρ_o is the resistivity at temperature T_o. The resistance may be written as

$$R_t = R_o\{1 + AT + BT^2 + C(T - 100)T^3\} \qquad (2.8)$$

for T from $-200°C$ to $+850°C$. The calibration must comply with the requirements of the International Practical Temperature Scale of 1968 (IPTS-68).

Figure 2.3 shows versions of the wirewound resistor which is considered as an industrial standard. Precision can be as good as 0.1°C.

Thermistor

The thermistor is a resistance thermometer made of semi-conductor material (a metal oxide). As the temperature rises so the resistance falls, unlike the resistance thermometer, and it varies much more than in metals. The resistance is also non-linear so that the temperature coefficient is given by

$$\alpha = -B/T^2 \qquad (2.9)$$

where T is in degrees Kelvin. Manufacturing tolerance can be as much as 20 percent and drift can take place with time. Thermistors with positive

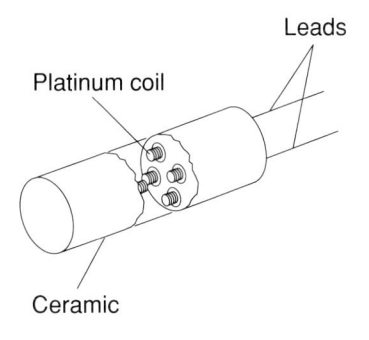

(a) Internally wound

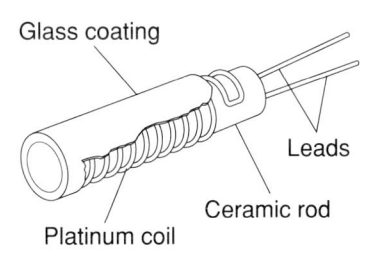

(b) Externally wound

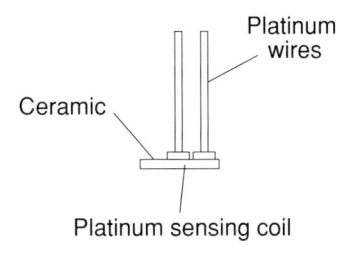

(c) Surface resistor

Fig. 2.3 Wirewound sensing element of platinum resistance thermometers (Courtesy of Fisher-Rosemount)

temperature characteristics are available. These are commonly used in series with electrical equipment to limit overheating as the resistance also increases rapidly above a fixed temperature.

Thermocouple

If two different metals are joined together and the junction is at a different temperature from the remainder of the circuit, then a voltage is generated which is proportional to the temperature. Ideally there should be two such junctions and the cold one should be kept at the ice point by being immersed in melting ice. There is a range of types for different applications (12). The problem of retaining the cold junction at 0°C is overcome in some industrial designs by allowing the junction to be at ambient temperature and compensating for this with some other device such as a resistance thermometer. In modern circuits with a high impedance amplifier the compensation is sometimes left until after amplification and the whole circuit is available in one chip. Base metal thermocouples are most common while the platinum/rhodium types offer a higher top temperature and are less susceptible to corrosion. Table 2.3 gives a selection of types.

Precious metal thermocouples can also be used down to 1K. Diagrams of a thermocouple circuit and a typical sensor design are given in Fig. 2.4.

Other types

Solid state devices are now available for temperature measurement. A silicon diode may be used since such devices are temperature sensitive. These may be incorporated in integrated circuits which provide a voltage or current proportional to the temperature.

Radiation measurement is used up to several thousand degrees.

Table 2.3 Thermocouple ranges (12)

Type	Wires	Typical Temperature range °C	Typical uncertainty
B	Platinum–rhodium	0 to +1700	±4°C
J	Iron–Constantan	0 to +850	±3°C
K	Ni-Cr/Ni-Al	−200 to +1200	±3°C
T	Copper/Constantan	0 to +400	±1°C

Commonly Measured Fluid Parameters

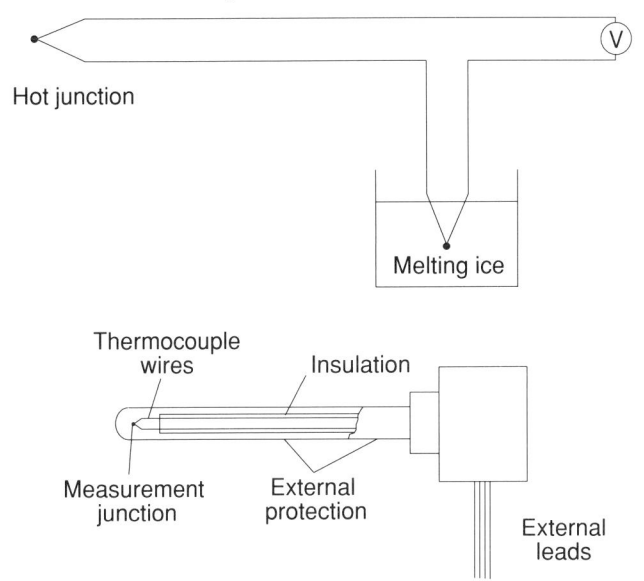

Fig. 2.4 Thermocouple circuit and diagram of sensor

2.2 PRESSURE

This section starts with a consideration of:

- pressure units and standard conditions;
- a qualitative description of Bernoulli's equation

Types of pressure gauge are then considered:

- manometer;
- mechanical devices which deflect under pressure;
- electro-mechanical pressure transducers.

This section concludes with a brief description of one means of calibration.

Pressure is the force per unit area which a body experiences if immersed in a gas or a liquid. The SI unit of pressure is the Pascal (Pa) which has the units of Newtons/metre² (N/m^2). This unit is too small for most uses and its size is well illustrated by Miller (**13**) who pointed out that an apple or a small glass of water exerts a force (against gravity) of roughly one Newton. If the glass of water were then to be poured over a tray of one square metre area, the pressure exerted by the water on the tray would be approximately one Pascal. For this reason the kilo Pascal (kPa), equal to 1000 Pascals, or the bar, equal to 10^5 Pascals and approximately one atmosphere, are more useful. Some useful pressure comparisons are given in Table 2.4.

Pressure acts equally in all directions. This is termed 'isotropic'. It is sometimes useful to refer to pressure as a 'head'. The relationship between head and pressure is easily derived for a certain depth of liquid. If a tube of liquid of cross-sectional area A and height h is held vertically, then the weight of liquid in the tube will be $\rho A h g$, where ρ is the density and g is the gravitational constant. The pressure on the bottom of the tube will then be

$$p_g = \rho A h g / A = \rho h g \tag{2.10}$$

Table 2.4 Atmospheric pressure and standard conditions

One atmosphere
 = 101 325 N/m²
 = 101 325 Pa
 = 101.325 kPa
 = 1.01 325 bar
 = 760 mm Mercury (Hg)
 = 10.3 m Water (H_2O)
 = 820 mm of water-over-mercury in a manometer
 = 14.71 psi (lb/in²)

Note also for gas correction:
– Metric standard reference conditions: 15°C and 1013.25 mbar dry
– Normal temperature and pressure: 0°C and 1013.25 mbar (SI equivalent values)

or rearranging, the head

$$h = \frac{p_g}{\rho g} \tag{2.11}$$

where p_g is the gauge pressure, the pressure above atmospheric. In other words the head in a still liquid with a free surface is the height of the surface above the point of measurement. These pressures are above atmospheric pressure and are known as gauge pressures. Absolute pressure is the pressure compared with a vacuum and differs from gauge pressure by the value of atmospheric pressure.

An equation which is frequently used in fluid engineering is that named after Bernoulli.

$$\frac{V^2}{2g} + \frac{p_g}{\rho g} + z = H \tag{2.12}$$

where z is the height of the point of measurement above a datum level, and H is the height of the still free liquid surface above the datum. This will be returned to in Section 3.7.

Figure 2.5 shows the use of the Bernoulli equation in obtaining the relationship between the variables for various points in the circuit, assuming that there are no losses (i.e., that the fluid is inviscid). The tank on the left is large enough so that the liquid can be taken as being still. The surface of the tank is the total head, H, above the datum. If one were to descend into the tank to a depth h, or a height z above the datum, a head h ($-H$ z) would be experienced, but since there will be negligible fluid movement, $V = 0$, and $h = p_g/\rho g$. The equation becomes

$$h + z = H \quad \text{or} \quad p_g + \rho g z = \rho g H \tag{2.13}$$

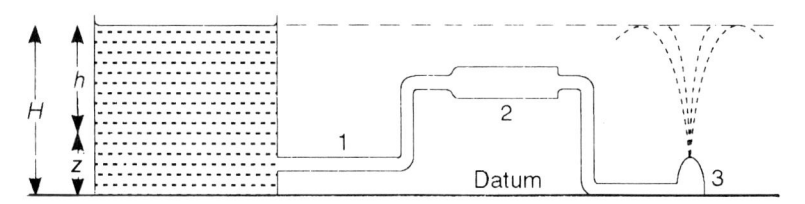

Fig. 2.5 Illustration of Bernoulli's equation

If one now were to swim into the tube at the outlet of the large tank at point 1, one would get carried along by a flow of velocity V and would experience a drop in pressure from that which was experienced in the tank at the same level. The pressure will be given by

$$p_{g1} = \rho g H - \rho g z_1 - \frac{\rho V_1^2}{2} \qquad (2.14)$$

where z_1 is now the height of the tube at 1 above datum. This pressure is known as the static pressure and it can be measured via a small hole in the pipe wall perpendicular to the direction of flow. If one could turn and face the flow, where it came to rest a greater pressure would be felt. This is the total pressure (cf equation (1.1)) given by

$$p_o = p + \frac{\rho V^2}{2} \qquad (2.15)$$

Moving on to Section 2 the velocity will drop, due to the greater cross-sectional area, causing an increase in the pressure. The tube height, however, is greater and this causes a drop in the pressure. The fountain in Section 3 has a surface which must sense the atmospheric pressure only, so that at all points in the fountain the pressure will be zero relative to atmospheric pressure. The velocity will, therefore, be given by

$$\frac{V_3^2}{2} = \rho g(H - z_3) \qquad (2.16)$$

and hence becomes zero when $z_3 = H$. Of course in practice there will be losses caused by friction in the flow throughout the system and so the fountain will not rise to the same height as the tank surface. The calculation of losses will be discussed in Chapter 4.

The measurement of pressure is an essential part of virtually all flow systems and is required in many flow devices and flowmetering installations. The measurement of static pressure in a pipeline requires a small hole to be made in the pipe wall and the pressure at that hole to be measured. This gives the value of the static pressure, provided the hole is cut cleanly and sharply, without burrs or chamfers, and is small enough. For instance the standard for differential pressure flowmeters (ISO 5167) requires, amongst other details, that the hole be circular, and the edges flush with the internal surface of the pipe wall and as sharp as possible.

The diameter should be less than $0.08D$, where D is the diameter of the pipe, and preferably less than 12 mm, and the tapping should be cylindrical for at least $2.5\times$ the diameter of the tapping.

The main methods used for pressure measurement are described below.

Manometer

In a manometer the height of a column of liquid provides a measure of the pressure in a liquid or gas. This depends on the head in a liquid, as explained above. The simplest type of manometer is a simple 'U' tube manometer (Fig. 2.6(a)) in which the pressure to be measured is connected to one side of the 'U' tube and the other is either left open to the atmosphere or is connected to a second pressure line to obtain a differential pressure measurement between the two lines. The pressure compared with atmospheric pressure is given by equation (2.11). It can be seen from this that while the height of the column provides the measurement variable, the sensitivity can be altered by changing the fluid to one of greater or less density. Thus a manometer using water (ρ_w = 1000 kg/m^3) will be 13.6 times more sensitive than one using mercury (ρ_m = 13600 kg/m^3). If the pressures are measured in water and the tubes connecting the pressure tapping to the manometer (impulse tubes) are filled with water and the manometer is filled with mercury, then the pressure is *not* obtained simply from the use of the height and the density of mercury. While a pressure difference results in a column of mercury in one limb of height h, in the other there will be a column of water also of height h, and the pressure difference will be the difference between the two. This gives a pressure difference of $(\rho_m - \rho_w)gh$ where $(\rho_m - \rho_w)$ is the difference in density between mercury and water. The precision of such a method will depend on several factors such as: constancy of tube bore; contamination of the bore causing changes in the meniscus of the liquid; precision of the scale; and discrimination in reading the levels of the liquid in the two limbs of the manometer.

In order to improve the precision of the manometer various designs have been used. Figure 2.6(b) shows a manometer with a large tank for the second limb. This has the advantage of being able to adjust one limb to a datum and then to make one measurement of the height of the other limb. This will require the positioning of the level of the liquid in the tank to be highly repeatable. In order to obtain more movement of the meniscus for a

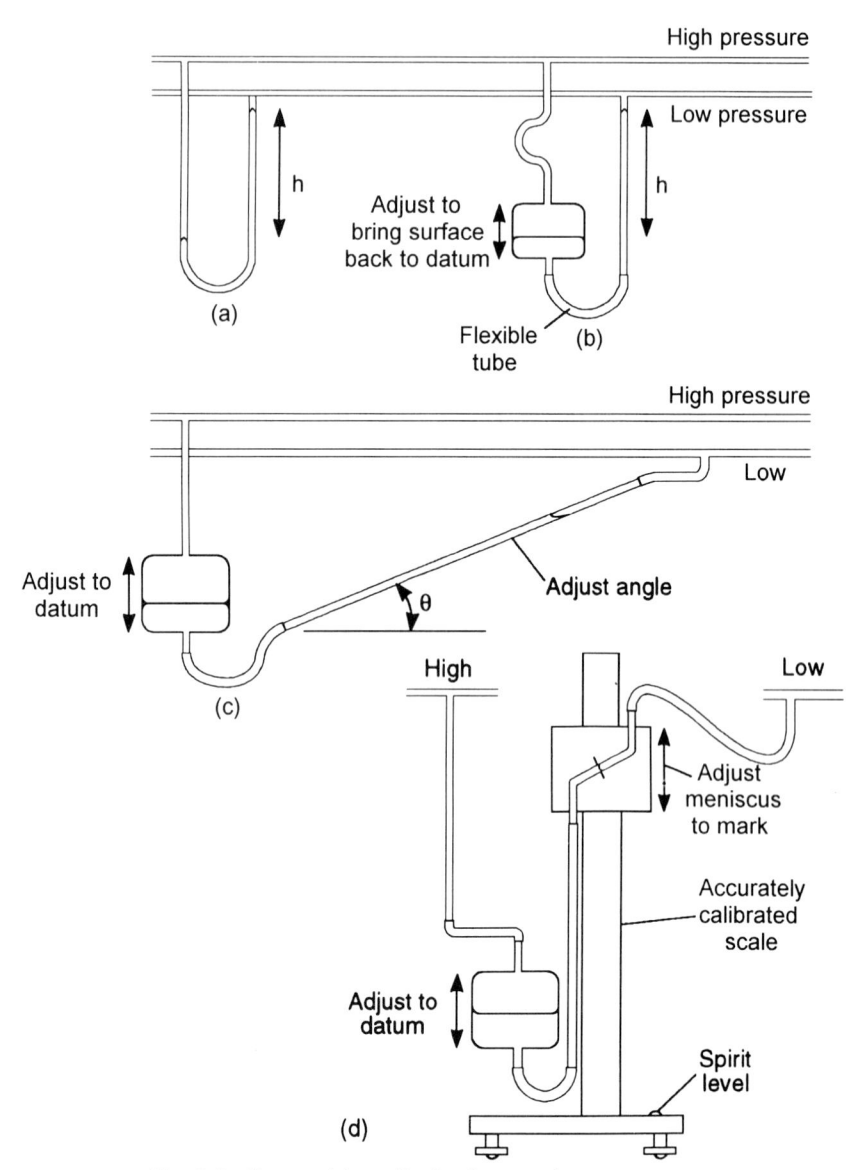

Fig. 2.6 Manometric methods of measuring pressure

given change in pressure, manometers have been designed with an inclined tube (Fig. 2.6(c)). The variation in the bore of the tube can cause problems in this design. To obtain the benefits of an inclined tube and yet to avoid the problems of variation in the bore, very precise manometers have been designed in which the inclined tube can be moved with a micrometer to return the fluid to the same point in the inclined tube (Fig. 2.6(d)). The reservoir is retained and is also adjustable since although the level is returned to a fixed point in one limb, the flexible tube connecting the limbs may change in volume as the position is adjusted and it may be necessary to set the reservoir level back to the datum level. This type of manometer can be used for greater deflection by combining a micrometer adjustment with a precision macro adjustment.

Manometers can also be mechanized by providing a magnet which floats on the mercury and a magnet follower which senses the position of the magnets and calculates the column length, or by other level-sensing methods.

Mechanical devices which deflect under pressure
One such device is the Bourdon tube which has a flattened tube wound into an arc. The tube unwinds under pressure since its flat cross-section becomes more oval and the increased distance between the arcs of inner and outer radii is accommodated by unwinding. The differential pressure version requires that the outside of the tube be pressurized. Figure 2.7(a) shows a typical example of such a device in which the tube end is connected in such a way as to move a pointer as the tube unwinds or to create an electrical signal change. For normal performance the tube is made from copper alloy or stainless steel. For high-precision instruments a low coefficient of expansion material may be used. The movement of the free end may be increased by using more turns in helical form (Fig. 2.7(b)) for high pressure or in spiral form (Fig. 2.7(c)) for low pressure (**14**). Sources of error in these devices are hysteresis, temperature effects, friction, and backlash. Ranges may be up to and in excess of 1,500 bar.

Another design uses diaphragms, as in Fig. 2.7(d). The diaphragm elements are made up from pairs of corrugated discs with spacing rings welded at the central hole. Pressure will cause the assembly to elongate and this will in turn be used as a method of registering the pressure change.

34   *An Introductory Guide to Industrial Flow*

(a) Bourdon tube

Movement with pressure change

Normal

High pressure

Cross-section

(b) Helical Bourdon tube

(c) Spiral Bourdon tube

(d) Diaphragm gauge

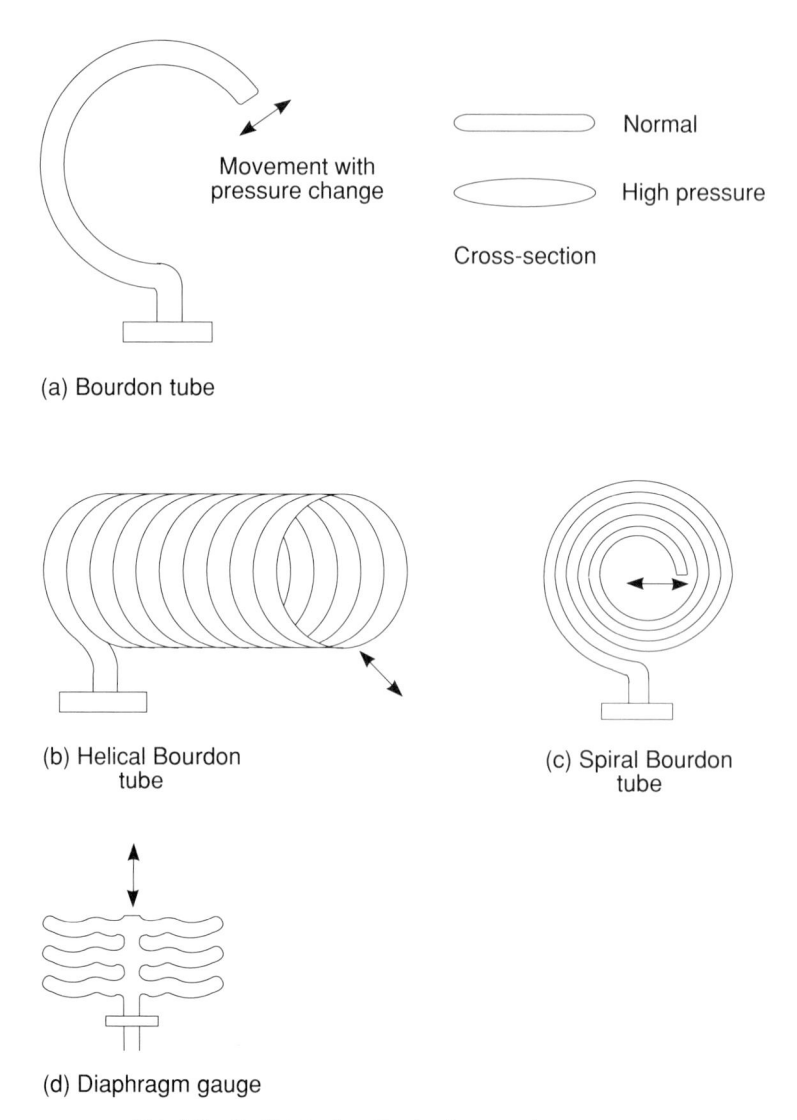

Fig. 2.7   Mechanical methods of measuring pressure

Electro-mechanical pressure transducers
This is the most important and precise device for the future and is in a state of continual development. Various methods have been used to obtain an electrical signal from a mechanical deflection.

(a) *Capacitance transducers* (Fig. 2.8(a)) in which deflection of one or two diaphragms causes movement of oil, which in turn deflects a third diaphragm which is a component in a capacitive element and the deflection is sensed as a change in capacitance.
(b) *Piezo resistive strain gauge* (Fig. 2.8(b)) in which the deflection creates a change in resistance of an element.
(c) *Piezo electric* (Fig. 2.8(c)) in which the pressure creates a voltage proportional to the pressure.
(d) *Resonant devices* (Fig. 2.8(d)) in which the principle of a violin string is used, whereby the string resonates at a higher frequency for higher tension. Thus if the tension on the string can be changed by a deflection due to pressure change, then the frequency can be related to pressure.

Calibration
Before leaving the subject of pressure measurement it is worth describing briefly the method of calibration using a dead-weight tester. Such a device is shown in Fig. 2.9 and consists of an hydraulic system with five terminal points:

(1) the screwed coupling for the pressure gauge which can be valved off;
(2) the loading piston which creates a known pressure in the system related to the weights on the piston;
(3) the pressurizing piston which is screwed in until the loading piston lifts off its seating;
(4) the filling point;
(5) the emptying point.

When the gauge is read it is common to spin the loading platform to ensure that friction in the piston, which might prevent movement, is already overcome, and thus that the load is transmitted with minimum friction loss into the hydraulic pressure throughout the system.

(a) Capacitance

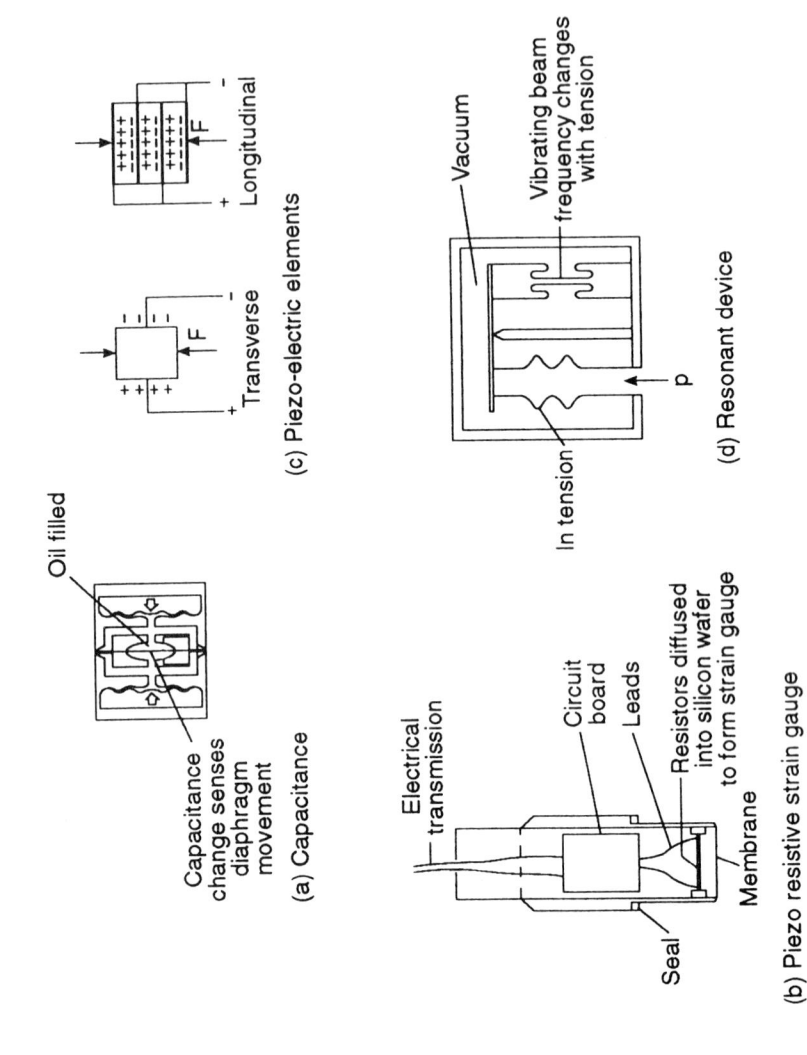

Capacitance change senses diaphragm movement — Oil filled

(c) Piezo-electric elements

Longitudinal / Transverse

(b) Piezo resistive strain gauge

Seal, Electrical transmission, Circuit board, Leads, Resistors diffused into silicon wafer to form strain gauge, Membrane

(d) Resonant device

Vacuum, Vibrating beam frequency changes with tension, In tension

Fig. 2.8 Electro-mechanical methods of measuring pressure (after Higham (14))

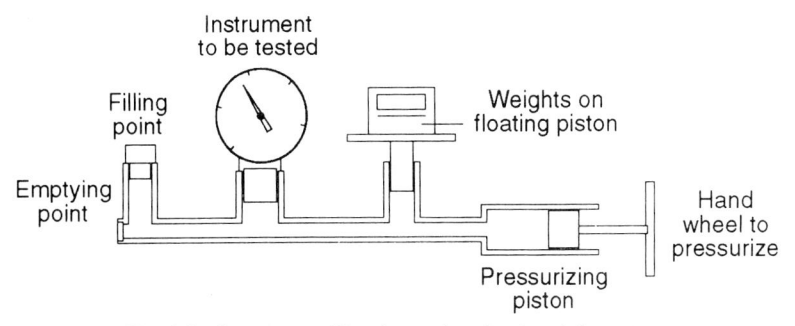

Fig. 2.9 Pressure calibration using dead-weight tester

2.3 DENSITY

> In this section density and related measurements are reviewed and some values given.
> The following types of measurement are described:
> - by volume and weight;
> - by oscillation/vibration;
> - by use of gamma radiation

The SI unit of density, ρ, is kilograms/metre^3 (kg/m^3). The density of water is approximately

$$1000 \text{ kg/m}^3 = 1 \text{ gm/cm}^3 = 62.4 \text{ lb/ft}^3$$

Specific volume is the reciprocal of density. Specific gravity is less commonly used today, but it is defined as the ratio of the density of a material to the density of water at 60°F. It has been replaced by Relative Density $(T_1/T_2 °\text{C or } °\text{F})$

$$= \frac{\text{Liquid density at } T_1°\text{C (°F)}}{\text{Density of water at } T_2°\text{C (°F)}}$$

The thermal expansion coefficient, β (sometimes α) – also known as the coefficient of volumetric or cubical expansion – is the fractional increase in specific volume, v, or decrease in density, ρ, due to an increase in temperature of 1°C

$$\beta = -\frac{1}{\rho}\frac{d\rho}{dT} = \frac{1}{v}\frac{dv}{dT} \qquad (2.17)$$

Water is an anomalous substance in that β is negative between 0°C and 4°C and positive above 4°C.

Compressibility for a liquid is the fractional decrease in specific volume or increase in density due to a change in pressure of 1 N/m²

$$= \frac{1}{\rho}\frac{d\rho}{dp} = -\frac{1}{v}\frac{dv}{dp} \qquad (2.18)$$

Compressibility is equal to the inverse of the bulk modulus, κ (Section 7.4). It should be remembered that for an ideal gas

$$\frac{p}{\rho} = pv = RT \qquad (2.5)$$

So for an ideal gas

$$\beta = \frac{1}{v}\frac{dv}{dT} = \frac{R}{pv} \quad \text{at constant pressure}$$

In Table 2.5 some typical values of density, specific volume, specific gravity (sg) and compressibility are given.

Table 2.5 Approximate values of density, specific volume, specific gravity and compressibility (values for gases at atmospheric pressure) (15)

	Temp °C	ρ kg/m³	v m³/kg	sg	Compressibility per N/m²
Mercury	20	13,600	0.0000735	13.6	3.8×10^{-11}
Water	15	1,000	0.001	1.0	4.9×10^{-10}
Benzene		700	0.0014	0.7	9.1×10^{-10}
Oxygen	0	1.43	0.70	0.0014	
Nitrogen	0	1.25	0.80	0.0013	
CO$_2$	0	1.98	0.51	0.0020	
Air	0	1.29	0.78	0.0013	

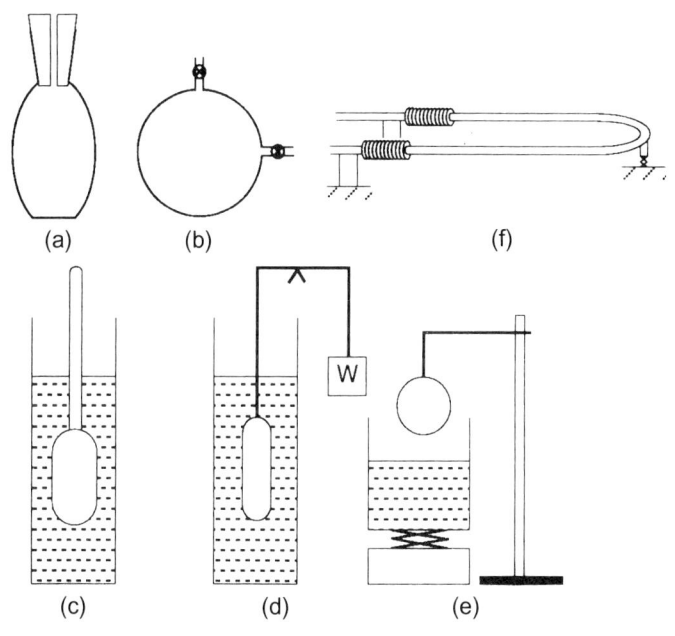

Fig. 2.10 (a) SG bottle pyknometer, (b) Pressure pyknometer, (c) Hydrometer, (d) Sinker, (e) Displacer, (f) 'U' tube weighing machine (after J. Stansfeld)

The density of a liquid can be calculated from a knowledge of the mass of a known volume of the liquid. A gas density may also be obtained in this way, but the sensitivity of the mass balance needed to obtain a precise value will be very high. Examples of sampling vessels and methods which use the upthrust on a submerged body (Chapter 3) are shown in Fig. 2.10.

Density can also be inferred from the difference in pressure between two different heights in a tank of the liquid. Several commercial instruments are available which make use of a change in natural frequency of vibration, either of an element immersed in the liquid or gas, or of a pipe through which the liquid or gas flows. Without going into the detailed mathematics or design of these devices the reader may find an explanation of the operation of these devices helpful. Referring to Fig. 2.11 the

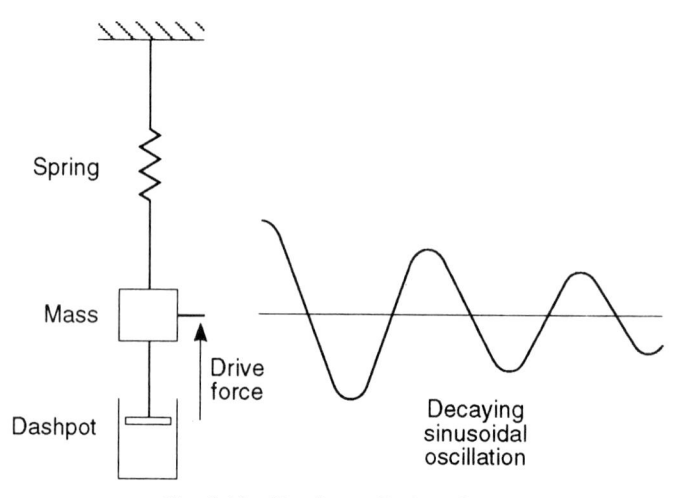

Fig. 2.11 Simple oscillation of mass

frequency of oscillation of the fluid container is governed by the equation

$$M\frac{d^2x}{dt^2} + \lambda\frac{dx}{dt} + kx = P_0 \sin \omega t \tag{2.19}$$

where M is the mass of the fluid container and fluid, λ is the damping constant, and k is the spring constant. A drive force, P_0, overcomes the decay due to damping and ensures that the oscillation takes place at resonance given by

$$\omega_r = \sqrt{\left(\frac{k}{M} - \frac{\lambda^2}{2M^2}\right)} \tag{2.20}$$

The density may be deduced from M if the mass of the vessel is known, the value of ω_r is measured, k is known, and the second term is negligible or known.

Outline diagrams of the essential features of commercial devices based on this principle are shown in Figs 2.12 and 2.13. The device for liquids (Fig. 2.12) offers a clear flow path and is little affected by viscosity and flow rate. The tube is isolated by flexible couplings. These instruments can have high precision.

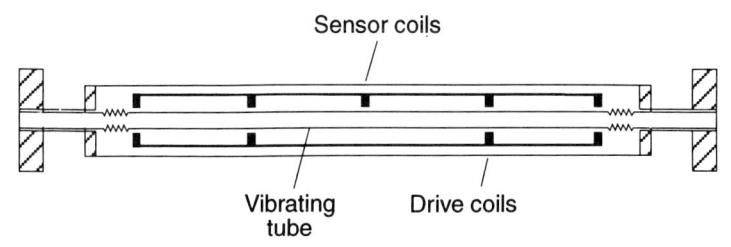

Fig. 2.12 Vibration density measurement: liquids (based on the Solartron device)

A resonant element gas density transmitter installed in a pipe is shown in Fig. 2.13. The cylinder vibrates in the hoop mode. The temperature coefficient is very low. The gas sample is taken from a tapping point in the line and flows around the cylinder before returning to the line. Filters are essential to retain performance. The head is inserted in the line to ensure that the temperature of the sensing region is the same as that of the

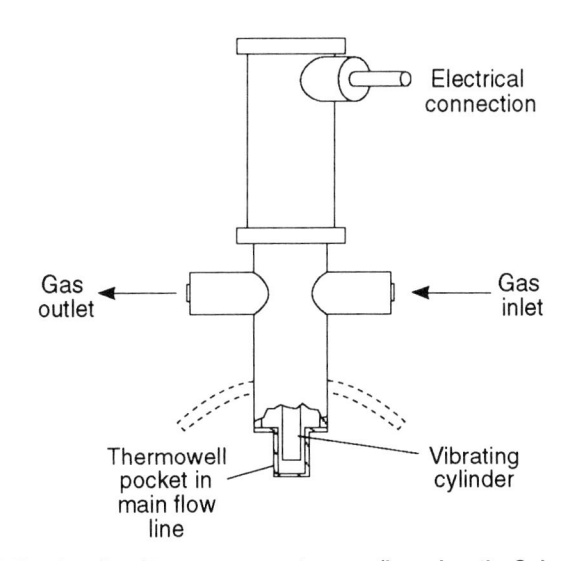

Fig. 2.13 Vibration density measurement: gases (based on the Solartron device)

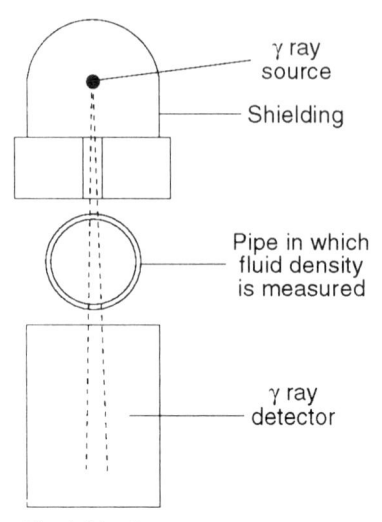

Fig. 2.14 Gamma ray density meter

gas. In some cases the element is inserted in the flow direct. Again this instrument has high precision.

Density may also be obtained from gamma ray and other nuclear radiations (Fig. 2.14). The radiation is attenuated by the quantity and size of the molecules in its path and obeys the following expression

$$I_o = I_i\, e^{-\mu\rho x} \tag{2.21}$$

where I_o is the emergent intensity of radiation, I_i is the incident intensity, μ is the mass absorption coefficient, ρ is the density and x is the distance traversed. The half-value (HV), or that thickness of a given material which will halve the intensity, is given by $HV = 0.69/\mu\rho$ mm. HV, for a source of about one megelectron volt, is about 100 mm for water and 15 mm for steel (**16**). This highlights the problem of measuring density of a liquid contained in a steel pipe. High precision and fast response are difficult to achieve.

For further information on this topic the reader is referred to more extensive references on the subject such as Higham (**14**).

Commonly Measured Fluid Parameters

2.4 VISCOSITY

Viscosity is recorded in two different ways:

Dynamic (absolute) viscosity, μ, is the ratio of shear stress to shear strain rate. The SI unit is the Pascal second (Pas), but in more common use is the centipoise (cP)

$$1 \text{ cP} = 10^{-3} \text{ Pas}$$

Kinematic viscosity, v, is the ratio of dynamic viscosity to density, μ/ρ. The SI unit is m^2/s, but in more common use is the centistoke (cSt)

$$1 \text{ cSt} = 10^{-6} \text{ m}^2/\text{s} = 1 \text{ mm}^2/\text{s}$$

Some typical values of viscosity are given below in Table 2.6.

In Fig. 2.15 it is noted that the viscosity of water decreases with temperature while the viscosities of air and steam increase with temperature at moderate pressures. The importance of common fluids, such as air and water, is that the value of viscosity is not dependent on the shear taking place in the flow. These fluids are referred to as Newtonian in their behaviour as compared with others where the viscosity is a function of the shear taking place. The behaviour of such fluids, known as non-Newtonian, is shown by the curves in Fig. 2.16.

Table 2.6 Approximate values of viscosity (Approximate values at 1 bar) (15)

	Temperature °C	Density kg/m^3 ρ	Viscosity Dynamic cP μ	Viscosity Kinematic cSt v
Water	20	1000	1.002	1.002
Benzene	20	700	0.647	0.92
Oxygen	0	1.43	0.019	13.3
Nitrogen	0	1.25	0.017	13.6
CO$_2$	0	1.98	0.014	7.1
Air	0	1.29	0.017	13.2

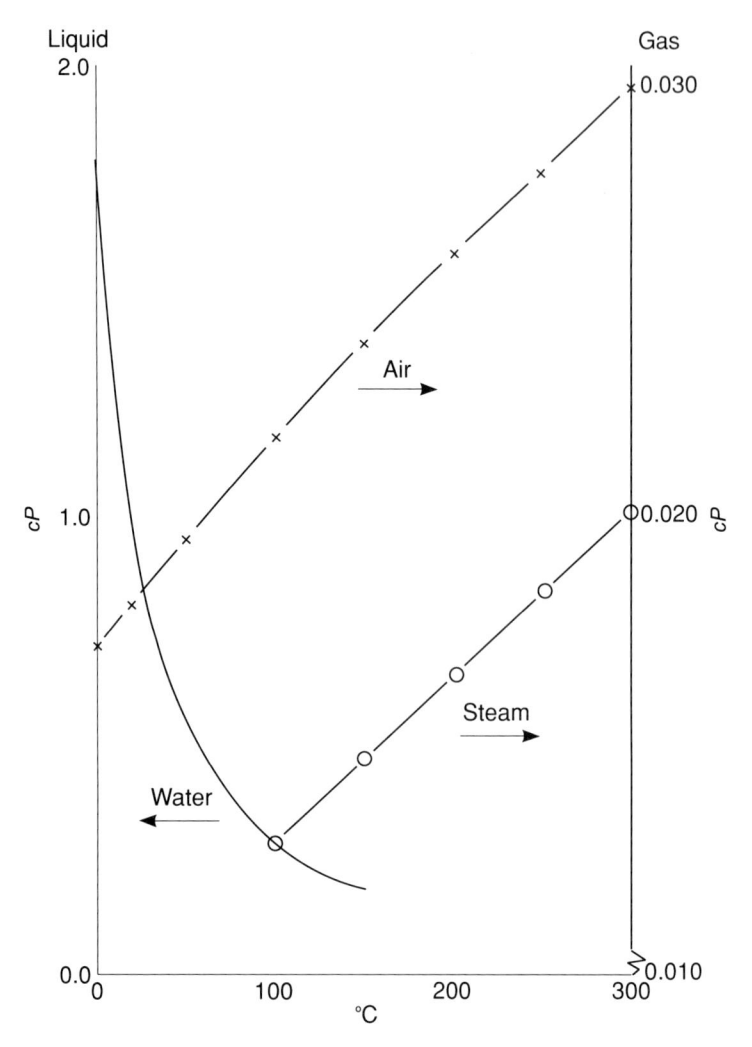

Fig. 2.15 Variation of viscosity of water, steam, and air with temperature at 1 bar

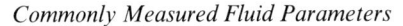

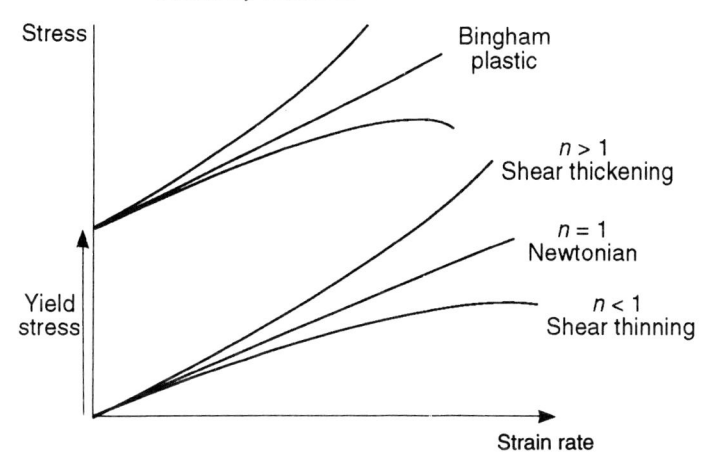

Fig. 2.16 Shear stress against shear strain rate for non-Newtonian fluids

---

Characteristic curves for non-Newtonian fluids:

Shear stress = K (shear strain rate)^n

For: $n < 1$ non-Newtonian fluid with shear thinning;
$n = 1$ Newtonian fluid;
$n > 1$ non-Newtonian fluid with shear thickening.

In addition some fluids exhibit a critical stress before flowing and others exhibit time-dependent effects. Obvious examples of this are the changes which occur in some foods when they are whisked, and indeed in food processing of milk it is important to avoid severe shear in the process as this may lead to the cream on the top of the bottles becoming too thick to pour. Another non-Newtonian fluid which may be affected by stirring is paint.

There are two main methods of measuring the viscosity of a liquid or a gas. The first uses the pressure loss in an orifice or a length of capillary under laminar flow conditions to obtain μ from equations which will be derived shortly (equations (3.8) and (3.9)) but which show that for a known and constant flow the viscosity is proportional to the pressure drop. The second method obtains the torque exerted by a thin layer of the liquid or gas in an annular cavity in which the central cylinder rotates and the outer wall is stationary. Other methods make use of a sphere falling or a bubble rising through the liquid. Walters and Jones give a fuller discussion of the types and equations which govern the behaviour of viscometers (**17**).

There are a variety of special measures of viscosity such as Saybolt seconds and Redwood. Information on these, as well as a useful range of viscosity data, can be found in Miller for many industrial fluids (**18**).

2.5 SURFACE TENSION

Liquids behave as if their free surfaces were perfectly flexible membranes having a constant tension σ per unit length which is called the surface tension (**19**). The value of the surface tension depends on various factors to do with the liquid and the liquid or gas on the other side of the surface, any surfactants which may be retained in the surface layer, temperature, etc. The surface tension is related to the fact that molecular attraction inwards into the liquid is not balanced by forces beyond the surface. Thus to distort the surface requires a force in part due to this tension.

Surface tension is in many cases negligible, but it has some important effects. In manometers, the surface level is raised due to surface tension for water, since the meniscus draws the water up the tube, but for mercury which has a domed meniscus, the surface tension lowers the surface. Surface tension is important in the break up of jets and in the break up of droplets in a two component flow, for instance of water in oil.

It appears to the author that small droplets of water can retain their spherical shape for some time when in motion on a horizontal metal surface, as if there is a time effect. However, this may be a result of contaminants on the surface.

2.6 CONCENTRATION MEASUREMENT

One very important liquid mixture is water in oil, as obtained from oil wells. The discussion in this section will focus on the measurement of the concentration of water in oil. The need for measurement is obvious, in that large amounts of oil are produced from oil wells, the oil is subject to tax, and the oil is sold. The producer does not want to be taxed on the water content and the purchaser does not want to buy water at the price of oil! It is, therefore, essential that the concentration of water in oil is known. In the North Sea in the early 1980s the concentration was only a few percent. As the wells have aged, the concentration has risen and the problem has changed. The most certain method of obtaining this quantity is to take a sample from a well-mixed pipeline flow and analyse it. However, it is not always straightforward to know whether the fluid in the pipeline is adequately mixed. Calculations can be made to obtain this mixing but they are at best approximate (**20**). Figure 2.17 shows the results of such calculations for a pipeline.

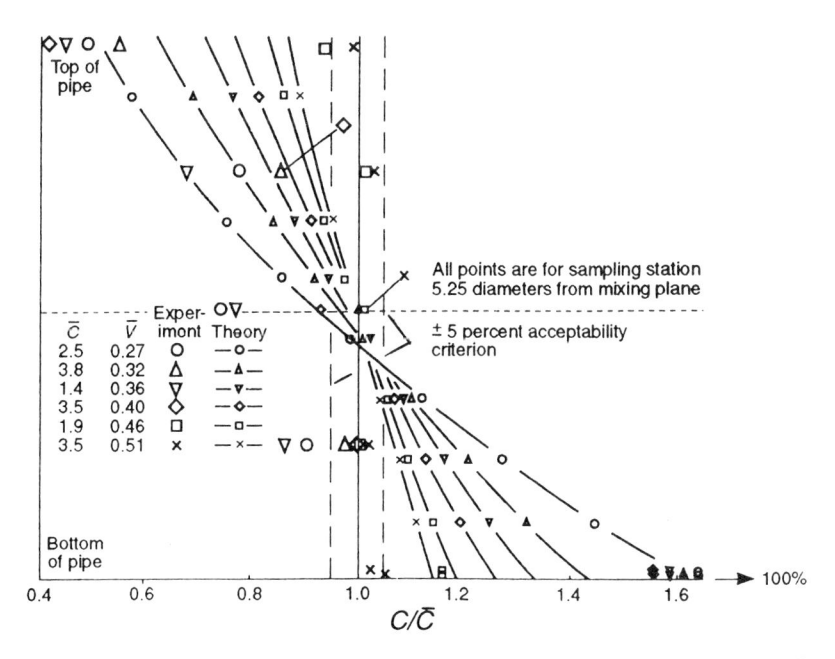

Fig. 2.17 Concentration variation across a pipe with water droplets in oil. Predictions are compared with a set of data supplied by BP ($\rho = 828$ kg/m^3, $v = 5.3$ cSt) (20)

Various methods of measuring the concentration of water in oil are used. Samples from a well-mixed pipeline where continuous extraction is taking place are often used. Capacitive methods are beginning to be used in which the electrical capacitance across the fluid is measured and from this the concentration of water in oil is deduced. The coriolis flowmeter can measure both mass flow rate and also density of the fluid, and it has been proposed as a means of measuring the concentration by comparing the density of the mixture with the known density of the components.

2.7 LEVEL

Methods of level measurement:

- float
- tank weight
- electrical
- ultrasonic
- microwave
- pressure

Methods to sense when a liquid surface passes a certain level:

- tilt switches
- electrical
- thermal
- optical
- ultrasonic
- gamma ray

The mechanical methods of measuring level are, in the main, straightforward. However, electrical techniques are also available. Methods include: sight glass, calibrated rod, float, tape and weight, microwave, thermal conductivity, capacitance, ultrasonic, load cell. Many of these are self explanatory, but brief descriptions and diagrams may be helpful in some cases.

Figure 2.18(a) shows a simple diagram of a float system. More sophisticated versions of this use a guide wire for the float and may use an RF surface sensor.

A load cell allows the weight of the vessel to be obtained and hence the volume and level. The electrical methods consist of an electrode (insulated

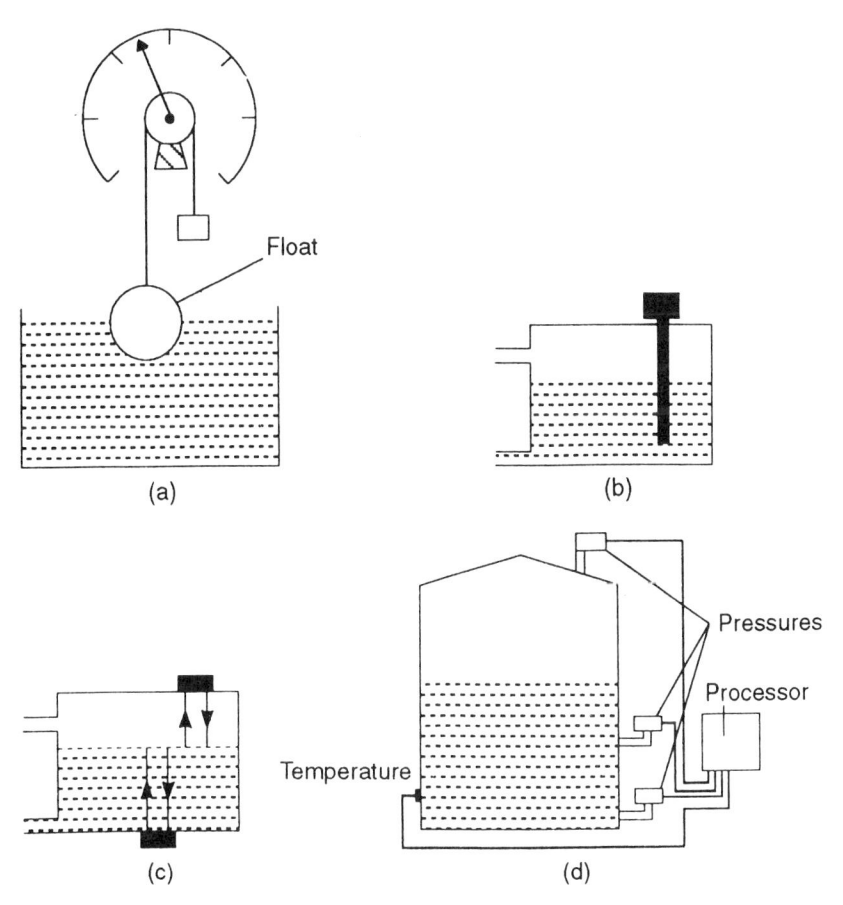

Fig. 2.18 Methods of measuring the surface level in a tank: (a) float method; (b) electrical sensing; (c) ultrasonic or microwave; (d) measurement of tank pressure

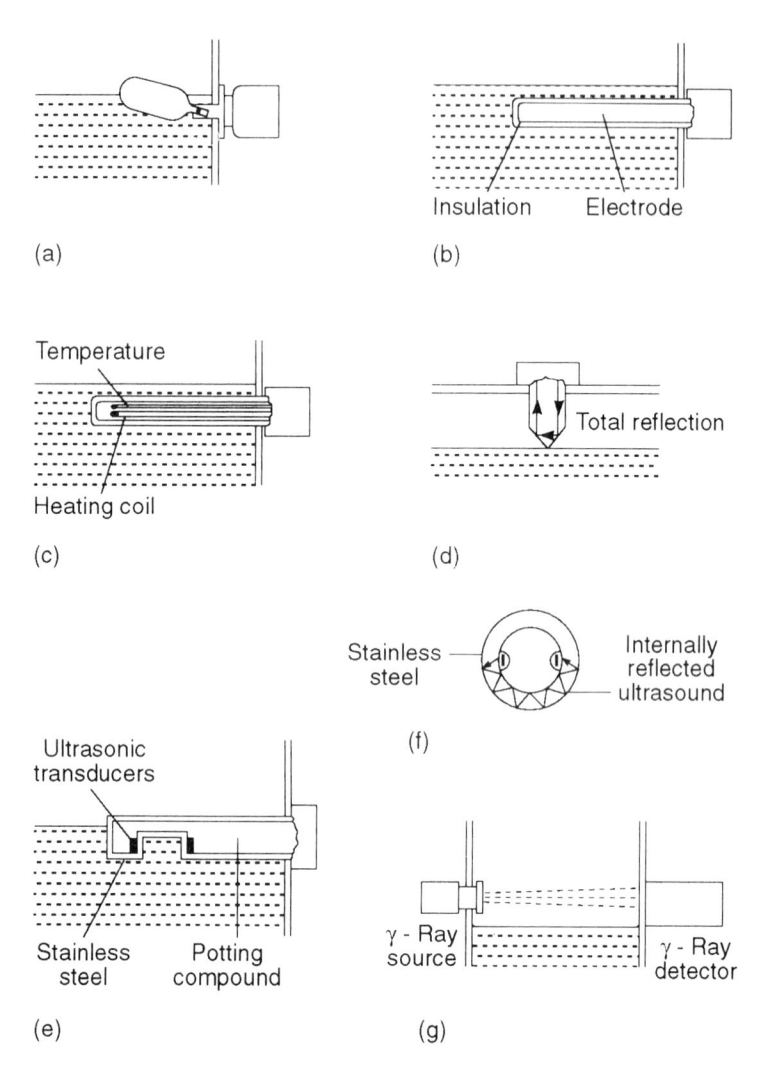

Fig. 2.19 Methods of measurement of the passing of the surface: (a) tilt switch; (b) electrical; (c) thermal; (d) optical; (e) ultrasonic transmission type; (f) ultrasonic reflection type; (g) γ-ray attenuation

for capacitance) which dips into the liquid (Fig. 2.18(b)) and the change in capacitance or resistance is calibrated against level change. Two electrodes can be used if the tank material is not suitable as a return path. The thermal systems depend on the heat transfer change when liquid surrounds a heated wire. This in turn will require a change in heating current and this parameter can be calibrated against level.

Sonic or ultrasonic pulses (Fig. 2.18(c)) or microwaves can be bounced off the surface either from above or below. The transit time of the pulse together with wave speed gives the height or depth.

The pressure at a point in the tank below the surface level can be interpreted as a level of liquid in the tank. Figure 2.18(d) shows a combined system for level, density, and stored mass for a closed vessel. One example of this uses a tube to the bottom of the tank through which air is bubbled. The back pressure in the line is then related to the pressure at the nozzle outlet and hence to the tank depth.

Other methods of sensing when the liquid surface passes a certain point depend on tilt switches, Fig. 2.19(a), the change in apparent mass due to the fluid inertia of a tuning fork or diaphragm and hence the change of resonant frequency, and the electrical, thermal and optical methods shown diagrammatically in Figs 2.19(b)–(d). For the optical probe, the light is fully reflected in air, but transmitted in liquid, and hence the received signal will indicate the moment when the liquid level reaches the probe.

Figures 2.19(e) and (f) show two methods using ultrasonic sensing. The first ultrasonic method makes use of the transmission and coupling change between liquid and gas, while the second uses the reduced reflection when the ring is immersed in liquid. Thus in the first a strong signal is created by the presence of the liquid and in the second a weak signal. The method in Fig. 2.19(g) depends on the attenuation when γ-rays traverse a denser medium.

CHAPTER 3

Basic Ideas of Fluid Mechanics

In this chapter we shall cover:

- the relation between pressure and depth in stationary liquid;
- the Reynolds number;
- the distribution of velocities in a pipe flow;
- the essential equations needed for this book culminating in Bernoulli's equation

3.1 HYDROSTATICS

Most of the contents of this book are concerned with fluid dynamics, the motion of liquids and gases. However, the situation in which they are stationary cannot be ignored. In Chapter 2 the variation in pressure in a tank below the level of the water surface was noted. If the atmospheric pressure is p_a and the water surface is at a height H above the datum from which water surface height is measured, then the pressure at a point z above the datum (h below the surface) will be given by

$$p = p_a + \rho g(H - z) = p_a + \rho g h \qquad (3.1)$$

This pressure acts equally in all directions. Imagine a submerged cubic object with two of its faces parallel with the surface (Fig. 3.1(a)). If the cube is positioned with the bottom surface z_1 above datum and the top surface z_2 above datum then the net upward force due to pressure on the cube will be

$$A(p_1 - p_2) = A\rho g(z_2 - z_1) \qquad (3.2)$$

But $A(z_2 - z_1)$ is the volume of the cube and so the upwards force on it is equal to the weight of the water which is displaced by it. This is a general rule, attributed to Archimedes, for any body submerged in a liquid. This force passes through the centre of volume of the displaced liquid. If the

54   *An Introductory Guide to Industrial Flow*

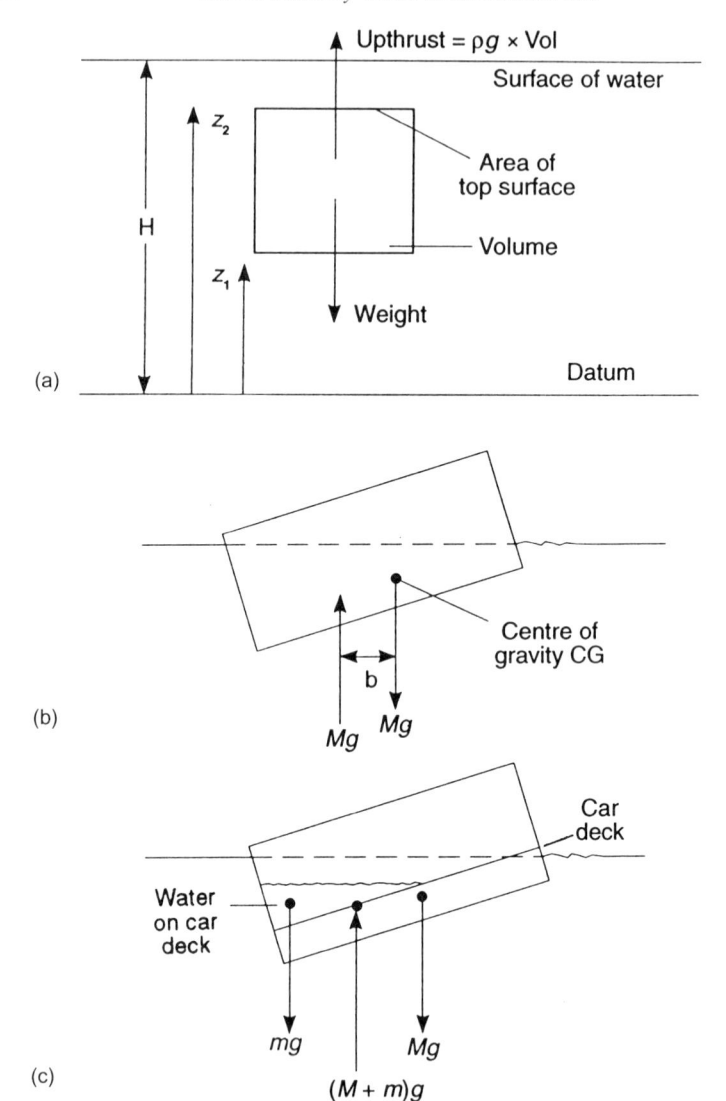

Fig. 3.1 (a) Upthrust on a submerged body
(b) Ship stability
(c) Ferry instability due to water on car deck

Basic Ideas of Fluid Mechanics

body is partially submerged (a ship) then the same rule applies. With simple shapes of body the centre is easy to identify. The upthrust through this point is equal to the weight of water displaced. The stability of a floating body may, therefore, be calculated from the positive righting moment. If the cross-section of the ship in Fig. 3.1(b) is approximately rectangular, and the ballast of the ship is contained in the bottom and constitutes most of the mass, M, then the righting moment is Mgb, where b is the distance between the vertical lines through the centre of gravity and the centre of buoyancy.

The effect of water collecting on the car deck of a roll-on-roll-off ferry can be seen in Fig. 3.1(c). The water will run to the lower side of the deck and the more the ship heels the greater the lever arm of this added load.

3.2 FLOW SIMILARITY

Reynolds number is a dimensionless quantity which determines the nature of the flow in pipes and machines

$$\text{Reynolds number} = Re = \frac{\rho VD}{\mu} = \frac{\text{Inertia forces}}{\text{Viscous forces}}$$

One of the most important concepts in fluid mechanics is similarity of flows. Many of the ideas in this book are concerned with the concept that flows in different pipes are similar in various ways, providing certain relationships are satisfied. In 1883 Osborne Reynolds set up an experiment to show this (Fig. 3.2).

In this experiment a dye streak was produced along the axis of the pipe through which the water flowed. The flow into the pipe was smooth, and for low flow rates the dye streak experienced very little change. However, although this behaviour could be sustained to high flow rates it was found that there was a critical condition below which the streak would always be well defined, but above which any disturbance in the flow would be magnified and the dye streak would be broken up by eddies.

The condition at which this occurs is given by the Reynolds number (Re) which provides an indication of the ratio of inertia forces in the flow

Reynolds' experiment

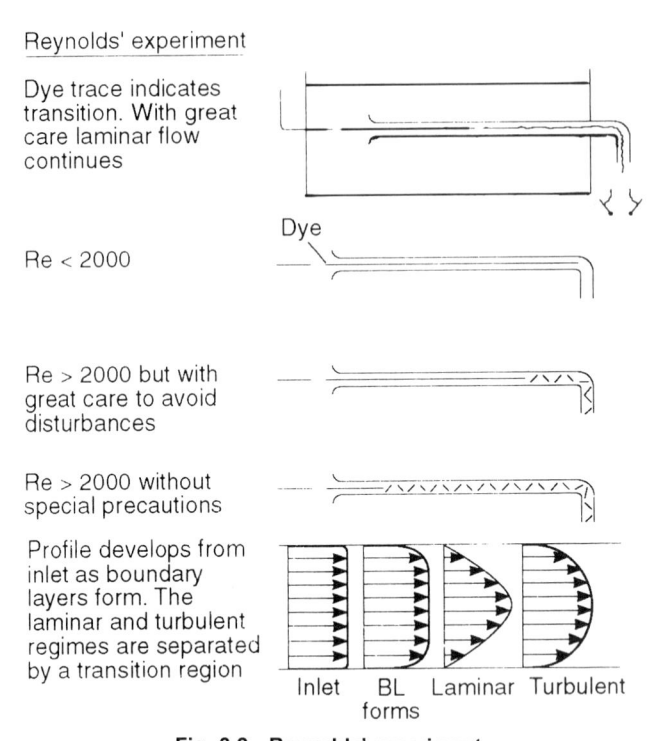

Dye trace indicates transition. With great care laminar flow continues

Re < 2000

Re > 2000 but with great care to avoid disturbances

Re > 2000 without special precautions

Profile develops from inlet as boundary layers form. The laminar and turbulent regimes are separated by a transition region

Inlet   BL forms   Laminar   Turbulent

Fig. 3.2   Reynolds' experiment

to viscous forces

$$Re = \frac{\rho V D}{\mu} \tag{3.3}$$

where ρ is the fluid density, μ is the dynamic viscosity, V is the velocity, D is the pipe diameter, and all quantities are measured in consistent units. For Re less than about 2000 the flow in the pipe is laminar and all the fluid travels in a direction parallel to the pipe axis (neglecting Brownian motion which leads to a slight blurring of the streaklines). It is shown below that the profile for laminar flow is parabolic. Above this value of Re it appears that there is an inherent instability in the flow which is triggered off by small disturbances and results in eddies which are of the same order of size

Basic Ideas of Fluid Mechanics

as the pipe. These eddies, which are superimposed on the axial flow, have a range of velocities up to about one-tenth of the axial velocity in the smooth pipe and a range of sizes. Their effect is to mix up the flow and to create a more uniform profile.

The disturbance may be due to vibration or due to disturbances already in the flow. It has also been shown that if great care is taken to eliminate all disturbances then the flow may remain laminar to values of Re much greater than 2000. However, for industrial flows the change to turbulent flow, the name given to the flow with eddies, occurs in the region of $Re = 2000$; this will be the normal state of industrial flows. This can be shown by taking typical values of the parameters. For water $\rho/\mu = 10^6$, $V = 1$ m/s and $D = 0.1$ m (4 ins), $Re = 10^5$. For air at ambient conditions $\rho/\mu = 0.7 \times 10^5$, $V = 10$ m/s and $D = 0.1$ m, $Re = 0.7 \times 10^5$.

The importance of flow similarity is that for flows in similar geometries, in this case pipes of circular cross-section, the flow pattern is the same, provided the Reynolds number is the same. Thus for different diameters, velocities, and fluids, the flow pattern can be the same in terms of the shape of the profile and the range and distribution of the eddies. In the case of similarity between the flows in two pipes there is an additional requirement that the roughness of the pipes is also in proportion to their diameters. Thus for a particular value of Re the flow pattern will be the same, regardless of the particular pipe diameter.

This similarity is sometimes a disadvantage. Film and television sequences which use scale models to simulate catastrophic happenings, such as fires, cannot match Reynolds numbers without using velocities out of proportion to the model. The result is that the turbulence patterns lack realism.

3.3 PIPE FLOW PROFILES

Why are velocity profiles important? The profile gives the distribution of velocity across the pipe. This pattern of velocity may have a profound effect on equipment or instrumentation which is in the pipe.

It has been found that for all continuous fluids the velocity of the fluid at the wall of the pipe is stationary relative to the pipe. This is the non-slip condition of fluid mechanics. Moving towards the centre of the duct, the

> **Non-slip condition**
>
> **Fully developed flow profiles:**
>
> – laminar profiles for Re < 2000
> – turbulent profiles for Re > 2000
>
> **Laminar flow is parallel with the pipe axis**
>
> **Turbulent flow has eddies mixing and flattening the profile**

velocity increases. But the actual pattern of the velocities in the pipe depends on the pipework upstream. If this is very long, a 'fully developed profile' is obtained, i.e. one which has reached an equilibrium shape. There is some argument about how long the pipe needs to be before this state is reached. Some put a figure of sixty diameters or so. There is evidence that in pipe flows the fully developed turbulent profile is not reached asymptotically, but rather that from a uniform inlet flow the profile becomes more peaked on the axis after about thirty diameters and may not have settled down even after seventy diameters (**21**). Twenty diameters may give a reasonable approximation to the shape, provided swirl is absent. This is a rotation of the flow which can be created by certain combinations of bends, valves, and other fittings. It is immediately apparent that industrial flows, due to bends and other fittings, seldom attain a fully developed profile.

There are various devices which will be affected by the shape of the profile. Flowmeters are obvious ones. They are carefully calibrated against a fully developed profile. The actual distorted profile may, therefore, change the calibration. But other devices may be affected by changed profile. A pump's efficiency may be reduced, and a fan may stall if subjected to poor flow conditions. Even the characteristics of some valves can be adversely affected by distorted profiles.

The losses which result from flow through ducts also need to be known, so that an assessment can be made of whether, for instance, a pump will have sufficient pressure and flow for a particular duty. This will be returned to in Chapter 4.

It is quite easy to show that the profile for a laminar flow in a pipe is parabolic. Referring to Fig. 3.3 the forces on the inner tube of fluid may be

Basic Ideas of Fluid Mechanics

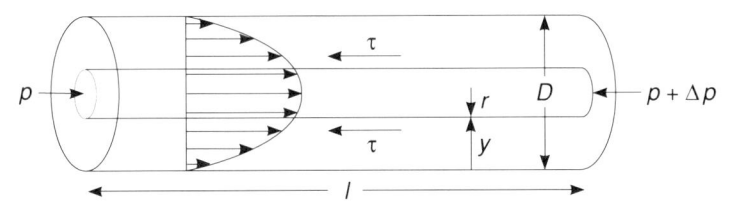

Fig. 3.3 Diagram to show force balance in laminar flow

balanced. On the outer surface of this the shear stress is given by

$$\tau = -\mu \frac{dV}{dr} \tag{3.4}$$

The force due to pressure difference on the inner tube over a length l is $\pi r^2 \Delta p$. Thus

$$\pi r^2 \Delta p = 2\pi r l \tau = -(2\pi r l \mu)\frac{dV}{dr} \tag{3.5}$$

Thus

$$\frac{dV}{dr} = -\frac{r \Delta p}{2\mu l} \tag{3.6}$$

and so

$$V = \frac{\Delta p}{2\mu l}\int_r^R r\,dr = \frac{1}{4}\frac{\Delta p}{\mu l}(R^2 - r^2) = V_o\left\{1 - \left(\frac{r}{R}\right)^2\right\} \tag{3.7}$$

where V_o is the velocity on the axis of the pipe and where V is zero at the wall of the pipe due to the non-slip condition for a fluid at a solid boundary. This gives a parabolic profile for V across the pipe. This is one example of Poiseuille flow, the other being for flow between parallel plates.

The mean velocity in the pipe is then given by

$$\bar{V} = \frac{V_o}{\pi R^2}\int_0^R 2\pi r\left\{1 - \left(\frac{r}{R}\right)^2\right\}dr$$
$$= \frac{V_o}{2} \tag{3.8}$$

where

$$V_o = \frac{1}{4}\frac{\Delta p R^2}{\mu l} \tag{3.9}$$

The fully developed profile as the fluid flows down a pipe is generated by the growth of the layer on the wall until a parabolic profile in fully developed laminar flow, as given above, is formed. If, however, the Reynolds number based on pipe diameter increases above about 2000 then instability will cause the flow to become turbulent as described above.

As soon as the flow becomes turbulent there ceases to be a simple relationship between shear and stress and the analysis of the profile becomes very complicated. For this reason a convenient expression for the turbulent pipe profile is

$$V/V_o = (1 - r/R)^{1/n} \tag{3.10}$$

where V_o is the centre line velocity, R is the pipe radius and n can be related to Re from experimental data (Table 3.1). This expression, therefore, not only provides a convenient representation of the boundary layer shape, but also an index adjustment to allow for the change in Reynolds number. The weakness of this curve is that it is not continuous at the centre line. The resulting profile shapes for the laminar and the turbulent regimes (based on equation (3.10)) are shown in Fig. 3.4.

Between the turbulent onset and the laminar there is a transition when the flow alternates randomly between laminar flow and patches of turbulent flow.

One interesting feature of these turbulent profiles is that the ratio of the velocity at about the 3/4 radius point to the mean velocity is approxi-

Table 3.1 Turbulent velocity profiles from equation 3.10

Re	4×10^3	2.3×10^4	1.1×10^5	1.1×10^6	2.0×10^6 to 3.2×10^6
n	6.0	6.6	7.0	8.8	10
$V/\bar{V}$ (at $r = 0.75R$)	1.0041	1.0045	1.0054	1.0055	1.0055
$V/\bar{V}$ (at $r = 0.758R$)	0.999	1.000	1.002	1.002	1.002

Laminar
Re < 2000

Turbulent
Re = 2.3 x 10⁴   Re = 1.1 x 10⁶   Re = 3.2 x 10⁶

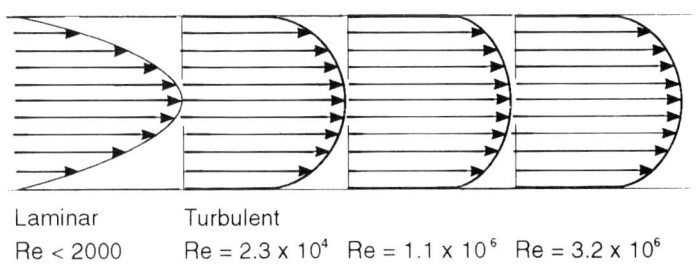

Fig. 3.4 **Laminar and turbulent pipe profiles**

mately unity. An expression for the mean velocity can be obtained by integrating equation (3.10).

$$\frac{\bar{V}}{V_0} = \frac{2n^2}{(n+1)(2n+1)} \tag{3.11}$$

Hence

$$V/\bar{V} = \frac{(n+1)(2n+1)}{2n^2}(1 - r/R)^{1/n} \tag{3.12}$$

and the values in Table 3.1 above are obtained. In fact if the radial position is taken as about 0.758 then the ratio is even closer to unity. This leads to the possibility of using a local velocity probe in a fully developed turbulent profile set at this point to deduce the mean velocity.

A final important point about the profiles in Fig. 3.4 is that while they are drawn as smooth curves, there is much difference between the flows. In the laminar profile the fluid shears smoothly over itself, but all moves essentially parallel to the axis of the pipe. In the turbulent profiles the curves represent a mean profile and ignore the turbulent eddies which can range up to as much as 10 percent of the velocity of the mean flow and cause fluctuations in all directions. In the context of flow measurement this is particularly significant in that the meter must measure precisely against this background noise.

3.4 CONTINUITY

To understand flow through pipe fittings and also to understand their effect on a pump network, the key equations which govern the distribution of flow and pressure need to be identified. These equations are the continuity and the energy equations. The first puts in mathematical form the idea of continuity, i.e. that in steady flow the same mass of fluid leaves a system as enters it, ensuring that mass is neither created nor destroyed.

Figure 3.5 shows an arbitrary duct which branches into two. The mass flowing through any section of the pipework is equal to the product of the cross-sectional area of the pipe, the velocity, and the density. Thus since the same mass must enter as leaves the pipework

mass flow entering at 1 = mass flow leaving at 2
   + mass flow leaving at 3

or written in terms of area, velocity and density

$$(q_m)_{\text{total}} = \rho_1 A_1 V_1 = \rho_2 A_2 V_2 + \rho_3 A_3 V_3 \tag{3.13}$$

If the density is constant then

$$A_1 V_1 = A_2 V_2 + A_3 V_3 \tag{3.14}$$

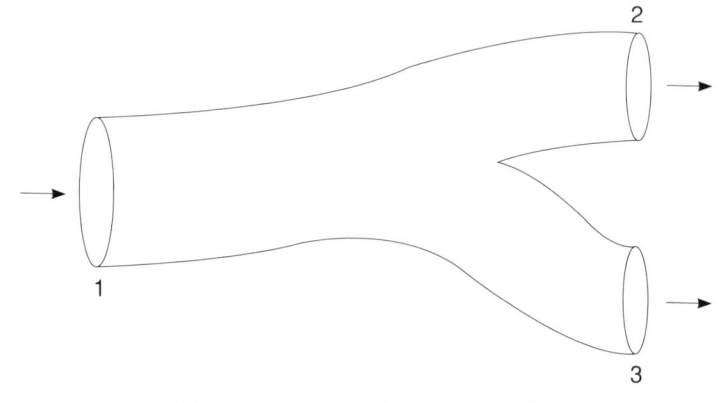

Fig. 3.5 Flow continuity through a branched pipe

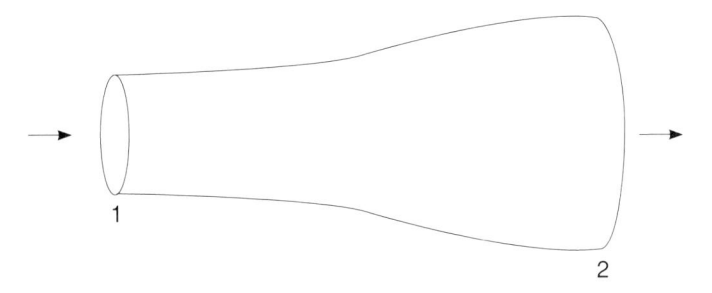

Fig. 3.6 Duct with varying cross section

Without further information there is an infinite number of possible flows. For instance if $V_2 = 0$ then

$$A_1 V_1 = A_3 V_3$$

whereas if $V_3 = 0$ then

$$A_1 V_1 = A_2 V_2$$

If we consider a single pipe in which the section increases and for which it is known that the density is constant, then

$$A_1 V_1 = A_2 V_2$$

or if, say, the exit diameter is twice the inlet diameter (Fig. 3.6) then the ratio of velocities is

$$V_1/V_2 = 4$$

However, this assumes that the velocity is constant across the pipe, an assumption which equations (3.7) and (3.10) show is incorrect, and the ratio can, therefore, only apply to the average values. If dealing with a gas in which the density is not constant (Chapter 6), under certain circumstances the velocity at Section 2 can actually be greater than at Section 1.

3.5 THE FIRST LAW OF THERMODYNAMICS AND THE ENERGY EQUATION

Without getting into a lengthy discussion of thermodynamics it is necessary to introduce some basic ideas. The first concerns **thermal** *equilibrium*. Two objects are said to be at the same temperature if no heat flows when one comes into contact with the other; they are said to be in 'thermal equilibrium'. The next concept is that of a *system*, by which the thermodynamicist means a definable region of matter, which can be distinguished from its surroundings by a well-defined boundary. The boundary need not be stationary and may be deformable, but must always enclose the same collection of matter. Most commonly this will refer to a fluid confined in a piston and cylinder or in a pipe or by the internal walls of a machine: a pump, compressor or turbine.

> **The first law of thermodynamics** implies that for a closed system:
>
> $$\frac{\text{heat added}}{\text{to the system}} - \frac{\text{work done}}{\text{by the system}} = \frac{\text{increase in the internal energy}}{\text{of the system}}$$

In symbols

$$Q - W = m(u_2 - u_1) \qquad (3.15)$$

where Q is the heat flow ***into*** the system, W is the work done ***by*** the system, m is the mass in the system, u_2 is the final internal energy of the system, and u_1 is the initial internal energy of the system.

It is easiest to understand this with a simple example. If some gas is confined in a piston with a cylinder then:

(a) if the gas is heated while keeping the piston stationary, the temperature will rise. Since for an ideal gas $pv = RT$ and the specific volume must be constant, the pressure rise will be proportional to the temperature rise relative to absolute zero. This increase in temperature and hence pressure is an indication of increased internal energy.

(b) if the piston is now released it will move so as to allow the volume to increase and the system – the trapped gas – to expand. The system has done some work, its pressure has decreased, and its internal energy has fallen.

This equation may now be extended by introducing the flowing system, the steady flow energy equation, and enthalpy (not to be confused with entropy).

The steady flow energy equation relates enthalpy, h, (defined in equation (3.16) below) velocity, V, and altitude, z, at the various points in a flow, to the exchange of heat and work occurring between those points. Of these quantities enthalpy needs further explanation. Figure 3.7 shows a device through which flow takes place steadily. Around part of this is a control volume which defines the volume of the device which is of interest to us. We allow a closed system to deform so that the small mass, δm, enters the control volume at 1 and an equal small mass, δm, leaves the control volume at 2. When the small element of fluid enters the control volume as the closed system deforms, work is done on the closed system. It acts like a small piston doing work. Similarly the small element of fluid leaving the control volume requires work to be done by the system. However, neither of these small bits of work are useful apart from moving the fluid into and out of the control volume. These work transfers always occur when elements of fluid enter the control volume and so in a flowing system it is convenient to combine them as enthalpy. Written in symbols this becomes (Fig. 3.7)

internal energy + work = enthalpy

so

$u\delta m + pv\delta m = h\delta m$

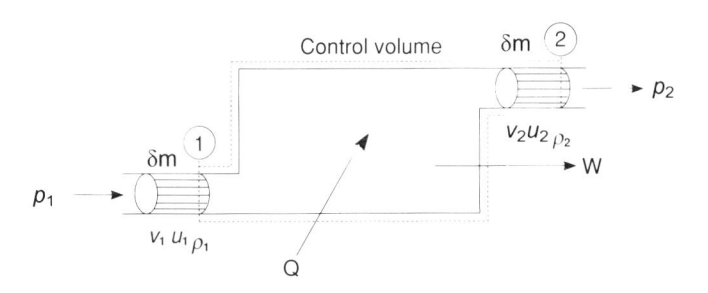

Fig. 3.7 Derivation of enthalpy for a flowing fluid

or

$(u + pv)\delta m = h\delta m$

Hence

$$h = u + pv \tag{3.16}$$

Since the most common fluid to which we shall apply this equation is a gas, it is useful to note two approximations which ease calculation.

- An *ideal gas* obeys $pv = RT$ (equation (2.5)) from which it is possible to deduce the fact that u is a function of T only (**22**) and hence that h is also a function of T only.
- A *perfect gas* obeys $u = a + bT$ (a and b being constants) from which it may be deduced that

$$u_2 - u_1 = c_v(T_2 - T_1) \tag{3.17}$$

and

$$h_2 - h_1 = c_p(T_2 - T_1) \tag{3.18}$$

and c_v and c_p are known as the specific heats at constant volume and pressure. It can also be shown that $R = c_p - c_v$, and the symbol, γ, is used for the ratio of the specific heats where $\gamma = c_p/c_v$.

For particular conditions and for greater accuracy the values of the internal energy, u, and the enthalpy, h, can be obtained from gas tables for given values of temperature and pressure. One common fluid which does not always approximate to a perfect gas is steam, and for this it is particularly useful to have access to empirical data. This book contains a part of a steam chart, known as a Mollier diagram. On this the values of enthalpy and other properties are marked for various conditions. As an alternative, steam tables can be obtained. Chapter 10 explains these in more detail.

The steady flow energy equation for the branched pipework in Fig. 3.5 can now be written as

Steady flow energy equation

heat added − work done = flow energy leaving
− flow energy entering

$$Q - W = q_{m_2}\left(h_2 + \frac{V_2^2}{2} + gz_2\right) + q_{m_3}\left(h_3 + \frac{V_3^2}{2} + gz_3\right)$$
$$- q_{m_1}\left(h_1 + \frac{V_1^2}{2} + gz_1\right) \qquad (3.19)$$

where Q and W are now heat and work transfer rates.

This can be applied to two examples:

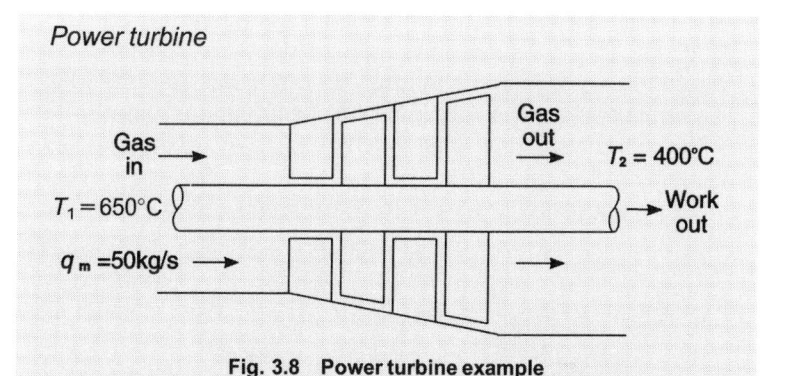

Power turbine

Fig. 3.8 Power turbine example

Referring to Fig. 3.8 the power extracted from the flow through a power turbine can be obtained assuming that the gas, although including some combustion products, approximates (Appendix A, p. 182) to a perfect gas. The inlet and outlet velocities are taken as negligible, and it is assumed that there is no heat transfer.

The inlet temperature of the gas is $T_1 = 650°C = 923K$, and the outlet temperature of the gas is $T_2 = 400°C = 673K$. The mass flow is $q_m = 50$ kg/s. Taking the value of c_p as 1.150 kJ/kgK, equation (3.19) gives (modified for one exit duct and omitting z terms since there is no change in z)

$$W = \frac{50 \times 1150(923 - 673)}{1\,000\,000} \text{ MW} = 14.4 \text{ MW}$$

Heated tube

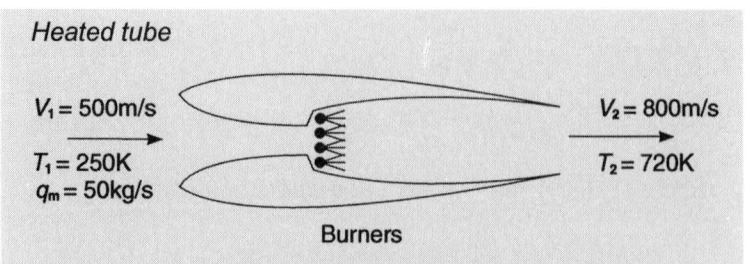

$V_1 = 500$ m/s
$T_1 = 250$ K
$q_m = 50$ kg/s

$V_2 = 800$ m/s
$T_2 = 720$ K

Burners

Fig. 3.9 Ram jet example

Referring to Fig. 3.9 the increase in temperature can be obtained for flow through a ram jet which is powering an aircraft flying at 500 m/s through air with temperature 250K. The outlet velocity is 800 m/s and the outlet temperature is 720K. If the mass flow is 50 kg/s find the heat added and the quantity of fuel needed.

Using equation (3.19) with $W = 0$, z terms omitted and one exit duct and approximating c_p as that for air.

$$Q = q_{m_2}\left(h_2 + \frac{V_2^2}{2}\right) - q_{m_1}\left(h_1 + \frac{V_1^2}{2}\right)$$

$$Q = 50\left(1006 \times 470 + \frac{640\,000}{2} - \frac{250\,000}{2}\right) = 33.4 \text{ MW}$$

If the heat of combustion of the fuel is 40 MJ/kg then the fuel flow required for this device would be 0.83 kg/s.

3.6 THE SECOND LAW OF THERMODYNAMICS AND REVERSIBILITY

The second law of thermodynamics may be stated as follows:

it is impossible to devise a cycle, the net effect of which is solely the absorption of heat and the production of work.

In a perpetual motion machine work comes from nothing. From the second law we have a new kind of impossible perpetual motion machine known as a perpetual motion machine of the second kind in which heat is wholly converted into work with no other changes taking place, through a series of heat and work exchanges between a fluid and the surroundings. For instance, a 100 percent efficient steam power-generating station, in which steam is heated in a boiler and work obtained through a turbine, is impossible by the second law. In order to develop this statement of the second law there is a trio of very important ideas.

- *adiabatic*
  This means that the system neither gains nor loses heat;
- *reversible*
  This means, literally, that the process can be reversed. The second law actually prohibits such a process, but nevertheless it proves a useful ideal which is approached by certain processes. The three primary causes of irreversibility are:

  - heat transfer across a finite temperature difference;
  - friction;
  - unrestrained expansion;

- *entropy*
  This is a property of a fluid. If that fluid undergoes a change which is reversible and adiabatic, then the process is said to be isentropic – of constant entropy. Entropy is given the symbol s (for unit mass), and we shall encounter it in the discussion of steam and the use of the steam chart. We shall also meet a highly irreversible process when we discuss shock waves in compressible flow.

The entropy of a system is given the symbol S, and the change in entropy of a system is defined as

$$S_2 - S_1 = \int_1^2 \left(\frac{dQ}{T}\right)_{REV} \tag{3.20}$$

where the subscript REV means that the integration takes place along a reversible path. For the present purposes one equation which includes entropy is introduced without derivation. (For a derivation see, for example, (**22**))

$$Tds = du + pdv \qquad (3.21)$$

Since

$$h = u + pv \qquad (3.16)$$

then

$$dh = du + pdv + vdp$$

and hence

$$Tds = dh - vdp \qquad (3.22)$$

which relates small changes in entropy, enthalpy, and pressure.

The concept of entropy is not straightforward and relates to the arrangement of particles in a volume and the statistics of rearranging them. In engineering its significance may be understood as follows.

(a) The area under the path of a reversible process on the temperature—entropy graph gives the heat exchanged (equation (3.20)).
(b) In an insulated system, where $Q = 0$, which is undergoing a reversible process, the entropy will remain constant (equation (3.20)). For instance, in equation (3.21), if $ds = 0$, then pdv, the work of a piston per unit mass, equals the change in internal energy.
(c) For a work transfer process which is carried out inefficiently, the entropy of the fluid will increase.

(These notes benefited from Haywood's presentation (**23**).)

Thus rewriting equation (3.22) for a perfect gas, since this is one of the most common fluids that we shall use, and using equation (3.18) in the form $dh = c_p dT$

$$s_2 - s_1 = c_p \int_1^2 \left(\frac{dT}{T}\right) - R \int_1^2 \left(\frac{dp}{p}\right) \qquad (3.23)$$

or

$$s_2 - s_1 = c_p \ln(T_2/T_1) - R \ln(p_2/p_1) \qquad (3.24)$$

Basic Ideas of Fluid Mechanics

Thus for a perfect gas (and as a good approximation to many gases) the change in entropy of the gas between two states 1 and 2, may be related to the ratio of the temperature and pressure changes using equation (3.24). For example, a perfect gas (approximating to air) with a value of c_p = 1.006 kJ/kgK and a value of R = 0.287 kJ/kgK, changing from 293K to 673K and from 1bar to 10bar will have a change in specific entropy of

$$s_2 - s_1 = 1.006 \ln(673/293) - R \ln(10/1) = 0.176 \text{ kJ/kgK}$$

This increase in entropy could indicate the inefficiency which occurred while compressing the gas adiabatically to about one quarter of its initial volume. On the other hand it could result from a two stage process of reversible compression followed by reversible heating.

3.7 DERIVATION OF BERNOULLI'S EQUATION

In Chapter 2 Bernoulli's equation was quoted (equation (2.12)). Here it is derived. If $Q = W = 0$, and we take one inlet and one outlet in equation (3.19), and we use equation (3.22) in the form for an isentropic process

$$\mathrm{d}h = v \mathrm{d}p \tag{3.25}$$

or

$$\mathrm{d}h = \mathrm{d}p/\rho \tag{3.26}$$

we obtain

Bernoulli's equation

$$\int_1^2 \frac{\mathrm{d}p}{\rho} + \frac{(V_2^2 - V_1^2)}{2} + g(z_2 - z_1) = 0 \tag{3.27}$$

This equation is known as the compressible fluid form of Bernoulli's equation. If ρ is constant for an incompressible fluid, then the Bernoulli equation (see equation (2.12)) can be rewritten as

$$\frac{p_2 - p_1}{\rho} + \frac{(V_2^2 - V_1^2)}{2} + g(z_2 - z_1) = 0 \tag{3.28}$$

An Introductory Guide to Industrial Flow

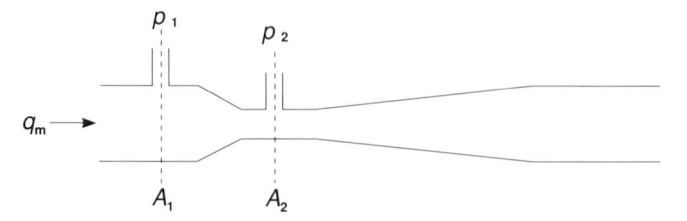

Fig. 3.10 Venturi meter

For cases where the third term in z can be neglected, a relationship between velocity and pressure drop can be obtained. It is found that

$$p_2 - p_1 + \rho\frac{(V_2^2 - V_1^2)}{2} = 0 \tag{3.29}$$

For the special case of a pitot tube $V_2 = 0$, and equation (1.1) is obtained with $k = 1$. Equation (3.29) can now be combined with the continuity equation, (3.13), rewritten as

$$V_2 = V_1 A_1/A_2 \tag{3.30}$$

to obtain

$$p_2 - p_1 + \frac{q_m^2}{2\rho}\left(\frac{1}{A_2^2} - \frac{1}{A_1^2}\right) = 0 \tag{3.31}$$

This equation relates the pressure change to the mass flow and is a very good approximation to the flow through a venturi meter as in Fig. 3.10. Suppose that the venturi meter is in a pipe of 50 mm diameter and has a throat diameter of 25 mm, then the mass flow of water for a pressure difference of 100 mm of water (from Table 2.4 equivalent to 984 N/m²) is

$$-984 + \frac{q_m^2}{2000}\left(\frac{1}{(0.000491)^2} - \frac{1}{(0.00196)^2}\right) = 0$$

So

$$q_m = 0.71 \text{ kg/s}$$

CHAPTER 4

Flow Losses in Pipes and Ducts

4.1 INTRODUCTION

To obtain the flow in a pipe system it is necessary to obtain the flow losses due to each component, in order to calculate the loss for the whole system.

In this chapter we consider a selection of flow components, the losses of each, and how to combine them.

For a full treatment the reader is referred to the book by D S Miller (**13**).

Bernoulli's equation allows us to relate energy in the flow at various points in the line, provided the flows are without loss. Bernoulli gives (equation (2.12))

$$\frac{p}{\rho} + \frac{V^2}{2} + gz = gH \tag{4.1}$$

where H is the total head and does not change in a reversible flow. It is, therefore, only valid for a flow without loss. Thus Bernoulli relates z, p, and V at each section of the flow. This equation can be used to obtain the velocity from the pressure drop measured, say, with a manometer, in a venturi. Such a flow is virtually lossless. However, we normally want to find the pressure drop caused by the losses in pipe components, and for these we use an equation obtained from dimensionless considerations and from experimental data.

4.2 LOSS COEFFICIENTS

In order to obtain the losses in a complete system it is necessary to know the loss caused by each system component, and this is provided by the head loss or pressure loss coefficient. The loss coefficient is not always constant with changing flow conditions and it has been found that a dimensionless group on which the value of the loss coefficient depends is the Reynolds number.

> **A pressure loss coefficient is introduced**
>
> $$K = \frac{\Delta p}{\rho V^2/2} = f(\text{geometry, inlet conditions, Mach number, fluid, Reynolds number, etc.}) \quad (4.2)$$

With loss coefficients of this sort an equation to relate the loss in a pipework system with the pressure across the system can be set up. We can also write this as a head loss coefficient

$$K = \frac{H}{V^2/2g} \quad (4.3)$$

Miller (13) has given a very thorough treatment of these losses in his book *Internal Flow Systems*, which is probably the most widely used industrial source book for such calculations. Suppose that we have a flow system consisting of three components in a pipe of constant cross-section so that V does not change. Suppose that the loss coefficients are K_1, K_2, and K_3. The total pressure loss will then be

$$\Delta p = \Delta p_1 + \Delta p_2 + \Delta p_3 \quad (4.4)$$

$$\Delta p = \tfrac{1}{2}\rho V^2 (K_1 + K_2 + K_3) \quad (4.5)$$

In a few cases the loss coefficient may be essentially constant. However, Miller shows that in most cases it varies with several parameters. Below are some examples of losses for common components and how these are combined for a simple system.

4.3 LOSS COEFFICIENT FOR A STRAIGHT PIPE

The equation for the loss in a straight pipe, as given by Miller, is

$$\Delta p = f_D \frac{L}{D} \rho \frac{V^2}{2} \quad (4.6)$$

Flow Losses in Pipes and Ducts 75

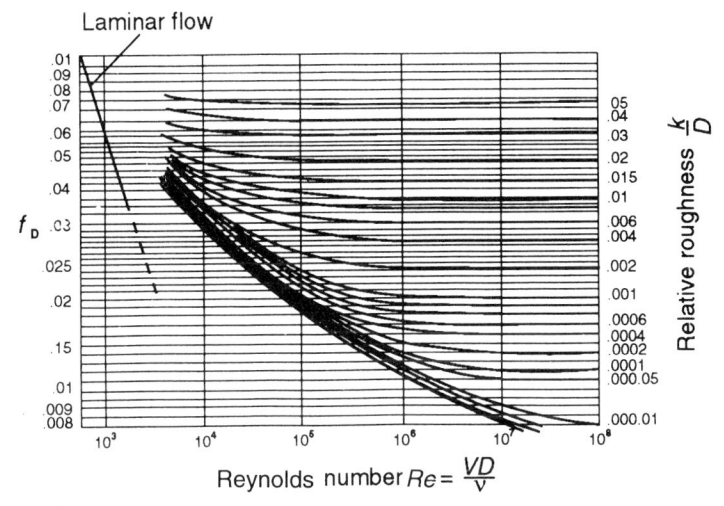

Fig. 4.1 The Moody chart for f_D (after Miller) (13)

The value of f_D can be obtained from the Moody chart shown in Fig. 4.1. The suffix D is used since this is sometimes known as the Darcy friction factor (24). In Chapter 6 and in the gas tables (Appendix A) f is used where $f_D = 4f$. f is sometimes known as the Fanning friction factor.

4.4 LOSS COEFFICIENT FOR INLETS

The loss coefficient for one type of inlet is shown in Fig. 4.2. In this case, as in the case of the other components below, the pressure loss is given by

$$\Delta p = \tfrac{1}{2}\rho V^2 K \tag{4.7}$$

4.5 LOSS COEFFICIENT FOR BENDS

In this book we shall concentrate on circular cross-section pipes and the simplest form of bends which can be described by the ratio of bend radius to pipe diameter, r/D, and the bend deflection angle θ. The loss coefficient, K, is shown in Fig. 4.3. Miller gives a correction factor for Reynolds

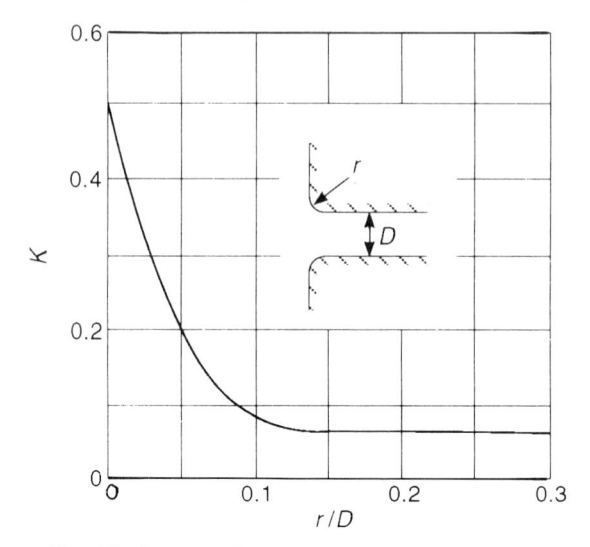

Fig. 4.2 Loss coefficient for inlet (after Miller) (13)

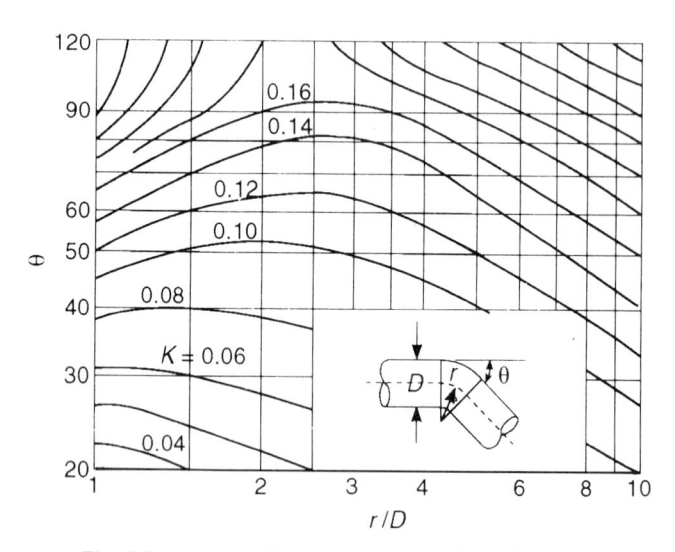

Fig. 4.3 Loss coefficient for a bend (after Miller) (13)

Flow Losses in Pipes and Ducts 77

numbers below 2×10^5 and a further correction factor for outlet straight length less than 30 diameters.

4.6 LOSS COEFFICIENT FOR VALVES

Miller provides the loss coefficient for various types of valve and one of these is reproduced in Fig. 4.4.

4.7 CALCULATION OF LOSSES FOR A SYSTEM

With the detailed information available from Miller (**13**) a precise value of the loss for each component can be found and these can then be summed to obtain the total for the system. In some cases this may be dependent on actual flow rate since some losses are Reynolds number dependent. In this case iteration may be necessary. In the following example the restricted loss coefficients given in this book are used to indicate the general method.

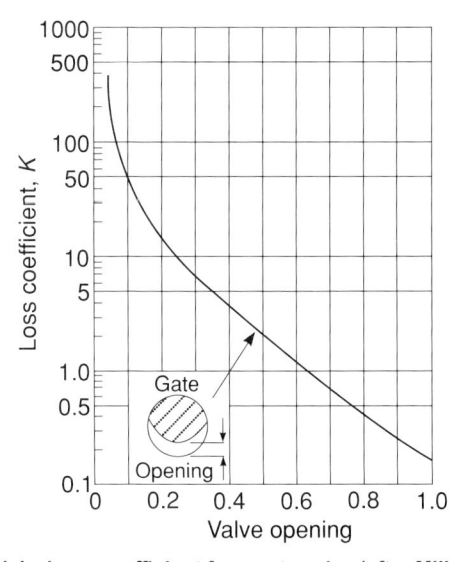

Fig. 4.4 Loss coefficient for a gate valve (after Miller) (13)

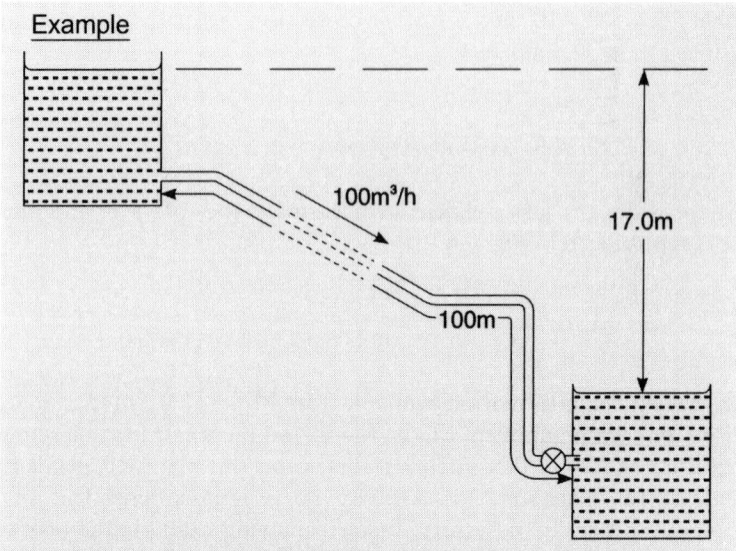

Fig. 4.5 Pipework configuration for example calculation

Obtain the pressure loss in the pipe configuration shown in Fig. 4.5 with a flow rate of 100 m^3/hr if the pipe is of 100 mm diameter.

Entry $r/D = 0.04$ $K = 0.25$ $\Delta p = \frac{1}{2}\rho V^2 K = \dfrac{1000 \times 3.54^2 \times 0.25}{2}$

$= 0.016$ bar

30° bend $r/D = 2$, $K = 0.065$ $\Delta p = 0.004$ bar

100 m pipe $k/D = 0.002$ $\Delta p = \dfrac{f_D L}{D}\dfrac{\rho V^2}{2} = \dfrac{0.024 \times 10^6 \times 3.54^2}{2}$
$Re = 3.5 \times 10^5$

$= 1.504$ bar

90° bends $r/D = 2$ $K = 0.16$ $\Delta p = 0.010$ bar each
Gate valve 0.6 open $K = 1.0$ $\Delta p = 0.063$ bar
Outlet – head is lost $\Delta p = 0.063$ bar

TOTAL PRESSURE LOSS   **$= 1.674$ bar**

$H = 17.0$ metres

For a full treatment of losses the reader is referred to Miller (**13**).

Flow Losses in Pipes and Ducts 79

4.8 FLOW CONDITIONING

As discussed in Chapter 3 the flow in a pipe may, in some applications, affect the performance of instruments or machines which are installed in the pipe. In such applications it is sometimes appropriate to modify the flow artificially. This may be done by using flow straighteners, flow conditioners, or a combination of both. Figure 4.6 shows diagrams of four long-standing devices, (a) the etoile (star) straightener; (b) the tube bundle straightener; (c) the perforated plate; and (d) the Zanker flow straightener. The first two aim to remove swirl from the flow, the third reorders the velocity profile and the fourth seeks to do both, that is to reorder the profile to approximate to a turbulent profile and to straighten the flow, getting rid of any swirl. Approximate diagrams are also given for the Mitsubishi flow conditioner and the K-lab flow conditioner, which are modern designs. It is important to remember certain points about these devices. Firstly, although they may remove most of the swirl and produce a profile which approximates to a turbulent profile, it is unlikely that they will ever recreate a turbulent profile with its shape and its turbulence distribution. Secondly, they will cause a pressure loss. This pressure loss may be obtained from equation (4.2)

$$\Delta p = K(\tfrac{1}{2}\rho V^2) \tag{4.8}$$

K is up to order 5 for the devices in Fig. 4.6 but for some flow conditioners may be as much as 15. The flow straightener or conditioner is most useful for ensuring that flowmeters are not subjected to a severely distorted flow. However, for high-precision measurements, the flow conditioner and flowmeter should be calibrated as one unit.

Bates (**25**) found strong swirl to be present in the gas metering system on a process platform in the Alwyn North field which had been installed to ISO 5167 in three 350 mm (14 inch) lines with orifice plates of orifice/pipe diameter ratio of 0.6. Thirty diameters separated the meter from a pipe reduction of 0.86 but upstream of this there were other fittings. Greasy deposits on the orifice plate gave an early indication of swirl being present. This was rectified by installing K-lab flow conditioners. This conditioner was machined from solid with holes (according to the diagram in the paper) at least $4d$ long on four radii and with a central hole. Bates claimed that field experience

a) Etoile

b) Tube bundle

c) Perforated plate

d) Zanker straightener

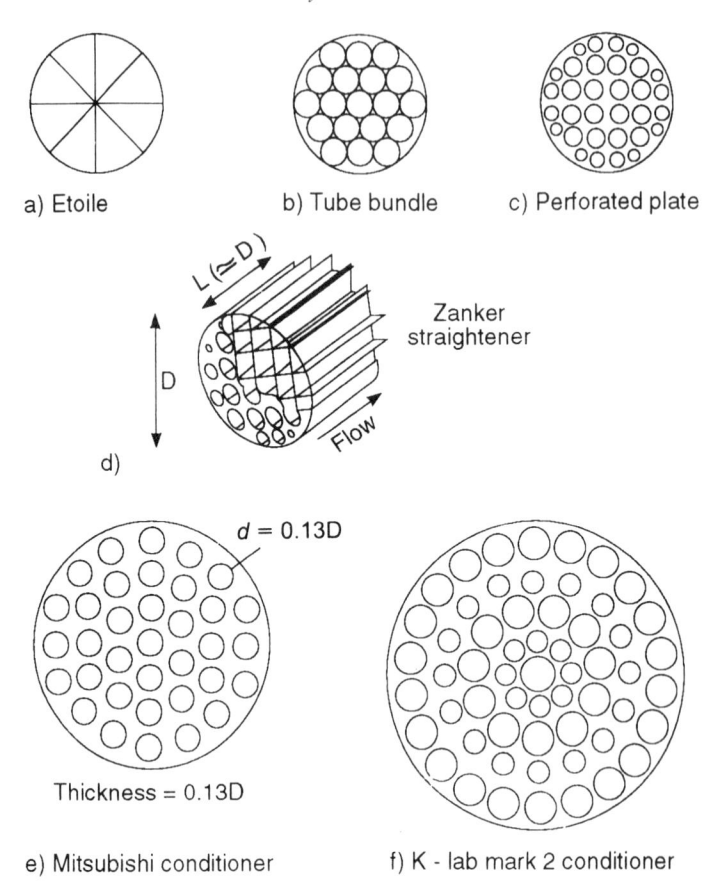

e) Mitsubishi conditioner (Thickness = 0.13D, d = 0.13D)

f) K - lab mark 2 conditioner

Fig. 4.6 A selection of flow straighteners and conditioners

had demonstrated that the flow conditioner is practical and cost effective.

Karrick *et al.* (**26**) addressed the important question of how closely the flow downstream of a flow conditioner represented fully developed flow as far as a flowmeter was concerned. Two rigs were used, one high and one low pressure, and measurements of profile and of orifice plate performance were made in each to determine the effect of an elbow or two elbows upstream of the tube bundle with various spacings between the (downstream) elbow, the flow straightener, and the metering position. They confirmed the point that, even though the profile may be approximately correct, the turbulence characteristics are unlikely to be.

CHAPTER 5

Flows in the Boundary Layer Next to a Duct Wall

- Boundary layer shape and dimensions
- Development of the boundary layer over a flat plate
- Thickness of laminar and turbulent layers on a flat plate
- Structure of the turbulent boundary layer
- Boundary layer separation
- Flow in small gaps

(detailed derivations will be found in Appendix D)

5.1 INTRODUCTION

An understanding of this topic is essential if the behaviour of flows and fluid machinery are to be understood. To allow the reader to follow the subject without too much mathematical detail, further development of the mathematics has been consigned to Appendix D.

There seemed to be an unbridgeable gap between experimental hydraulics and theoretical hydrodynamics until Prandtl (**27**) produced his theory (1904) of the boundary layer. One example of the gap was the observed ability of a body to produce lift and the inability of the theory to predict this. Boundary layer theory was able to reconcile these conflicting positions. Figure 5.1 gives a sketch of the boundary layer shape and dimensions.

Some idea of the development of the boundary layer profile as it flows over a flat plate may be obtained from Fig. 5.2.

5.2 DEVELOPMENT OF THE BOUNDARY LAYER

It is useful to consider some approximate calculations to obtain order of magnitude values for the main velocities associated with the boundary

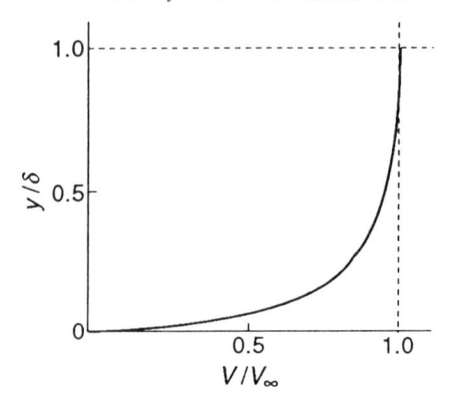

Fig. 5.1 Boundary layer shape and dimensions

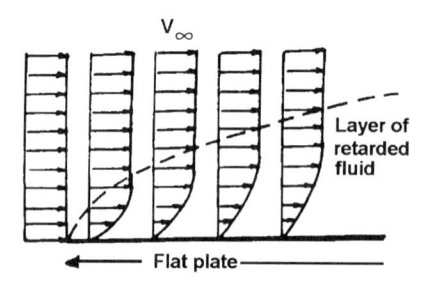

Fig. 5.2 Development of the boundary layer as it passes over a flat plate (after Prandtl) (27)

layer and its dimensions. If a simple, and highly approximate, boundary layer shape is taken (see Fig. 5.3 – the shaded area shows the velocity decrement from a uniform flow), it can be seen that the flow parallel to the wall entering the control surface is greater than that leaving the control surface. Therefore, there must be a component of flow leaving the control surface perpendicular to the wall. In Appendix D a simple calculation shows that the velocity perpendicular to the wall at the boundary of the layer is of order $\delta V_\infty / l$ where δ is the boundary layer thickness, V_∞ is the

Flows in the Boundary Layer Next to a Duct Wall

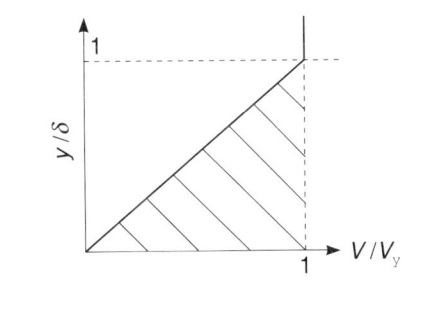

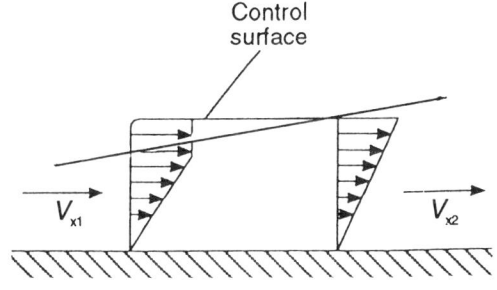

Fig. 5.3 Linear boundary layer development (19)

free stream velocity, and l is the distance from the leading edge of the plate. The thickness of the boundary layer can also be obtained as $\delta^2 \approx vl/V_\infty$.

For flow over a flat plate the boundary layer will initially be laminar and after a transition region where the nature of the layer alternates between laminar and turbulent flow, it becomes turbulent. Mixing is much more vigorous in the turbulent region and, as a result, the growth of the layer is faster. The thickness of the layers can be seen in Fig. 5.4 which takes a case for which both laminar and turbulent layers could just exist. The growth of a laminar boundary layer is shown in Appendix D to be $\delta = 4.99 \, (vx/V_\infty)^{1/2}$ and that for turbulent is given by Kay (**28**) as $\delta = 0.379x \, (v/V_\infty x)^{1/5}$. It can be seen that for water at 0.4 m/s after 2.5 m the thickness of the laminar layer is 0.012 m (12 mm) and the thickness of the turbulent layer is 0.06 m (6 cm). However, transition usually occurs for a Reynolds number between 10^5 and 2×10^6. The shape of the laminar and the turbulent layers are compared in Fig. 5.5. The violent mixing

For Re = 10^6

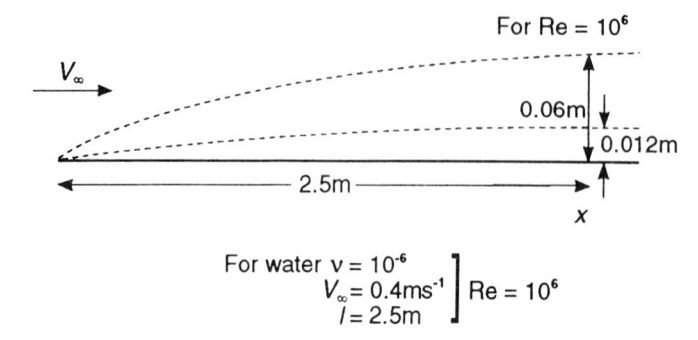

0.06m
0.012m
2.5m
x
V_∞

For water $v = 10^{-6}$
$V_\infty = 0.4 \text{ms}^{-1}$
$l = 2.5$m
Re = 10^6

Fig. 5.4 Thickness of laminar and turbulent layers on a flat plate

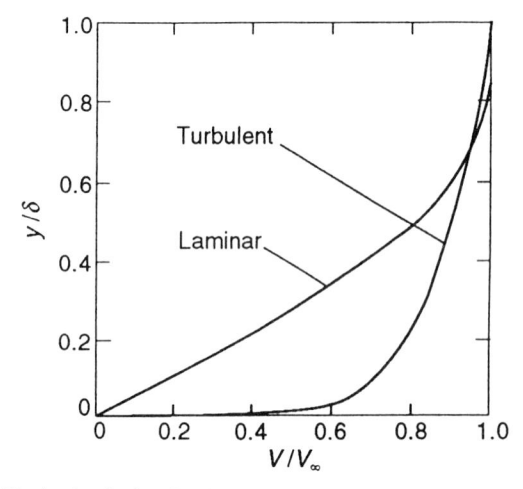

Fig. 5.5 Typical velocity distributions on a flat plate at zero incidence (19)

which occurs in the case of the turbulent layer results in a much more uniform velocity until close proximity to the wall. A useful power law (similar to that used for turbulent pipe flow profiles) which gives a reasonable fit for the experimental data of the turbulent profile is due to Nikuradse.

Nikuradse's power law for turbulent profiles

$$\frac{V}{V_\infty} = \left(\frac{y}{\delta}\right)^{1/7}$$

where V is the velocity at y.

There does not appear to be any theoretical basis for Nikuradse's curve. Figure 5.6 compares it with a curve from experimental data.

5.3 STRUCTURE OF THE TURBULENT BOUNDARY LAYER

The layers of the turbulent profile are shown diagrammatically in Fig. 5.7. The outer region consists of a moving pattern of large eddies interspersed with regions of relatively calm flow. To a moving observer, the edge of the boundary layer looks very irregular and consists of peaks of turbulence with valleys of still fluid in between. To the stationary observer (e.g. the

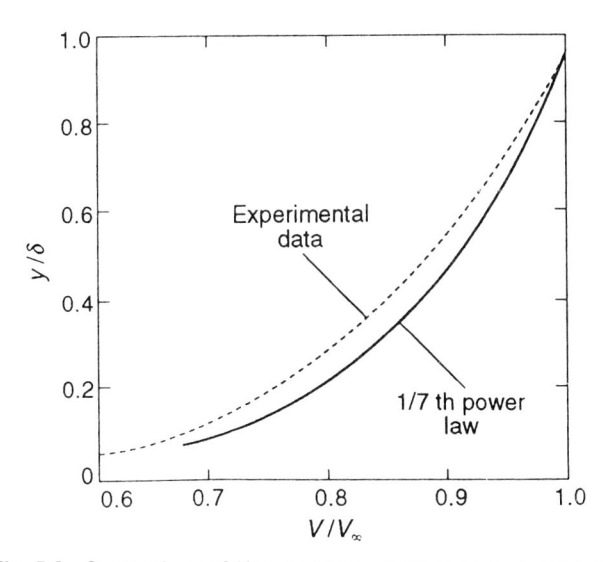

Fig. 5.6 Comparison of 1/7th power law with experimental data (19)

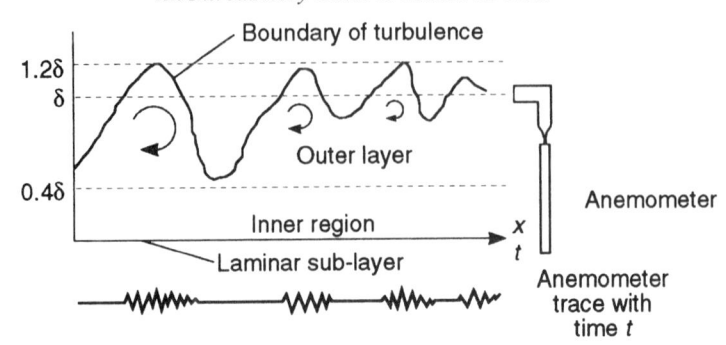

```
                    Boundary of turbulence
1.2δ  ┄┄┄┄┄┄┄┄┄┄┄┄┄┄┄/┄┄┄┄┄┄┄┄┄┄┄┄┄┄┄┄┄┄┄
  δ   ┄┄┄┄┄┄┄┄┄┄┄┄┄┄┄┄┄┄┄┄┄┄┄┄┄┄┄┄┄┄┄┄┄┄┄
                          Outer layer
0.4δ  ┄┄┄┄┄┄┄┄┄┄┄┄┄┄┄┄┄┄┄┄┄┄┄┄┄┄┄┄┄┄┄┄┄┄┄          Anemometer
                    Inner region         x
                    Laminar sub-layer    t
                                            Anemometer
                                            trace with
                                            time t
```

Fig. 5.7 Diagram of turbulent boundary layer

anemometer) the turbulence will appear spasmodic. The outer edge of the turbulence stretches to about 1.2δ. In this outer region the eddies are predominantly large. The fully turbulent region extends from 0.4δ down to the laminar sub-layer, with a wide range of turbulence frequencies and eddy sizes. The large low frequency eddies near the surface of this layer extract energy which is then dissipated in the smaller eddies farther into the boundary layer as heat. In the sub-layer the turbulent fluctuations become small, and viscous shear stresses predominate (**19**).

The gustiness of a strong wind will be largely due to turbulent eddies. Typical terrestrial boundary layer thicknesses are as much as 600m. Beyond this is the geostrophic wind. Near the earth's surface the wind speed is typically a half to two thirds of the geostrophic wind speed, which tallies with Fig. 5.5 for a turbulent boundary layer (**29**).

It should be noted that in the free stream of a turbulent flow, but away from the boundary layer, the turbulence tends to isotropy – the same in every direction.

In convergent and divergent ducts the boundary layer experiences favourable or adverse pressure gradients (decreasing or increasing pressure). This results in the fluid near the wall being accelerated or decelerated. This in turn leads to a thinning or a thickening of the boundary layer, as shown in Fig. 5.8. Nikuradse observed the first occurrence of separation at a semi-angle between 4.8 degrees and 5.1 degrees. For a description of the transitory stall which occurs in a diffuser and other flow regimes the reader is referred to Ward-Smith (**24**). The random attachment of the flow

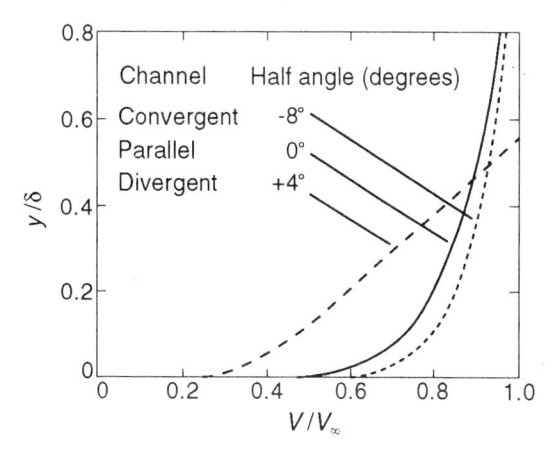

Fig. 5.8 Boundary layers in convergent and divergent ducts (after Nikuradse)

in a certain regime is used with feedback in the fluidic flowmeter, and in other oscillating devices.

Boundary layer separation is shown in Fig. 5.9. Separation is a result of the boundary layer encountering an adverse pressure gradient. The lower momentum fluid near the wall is slowed down until it comes to rest and this is followed by a reverse flow. The diagram shows this in an exaggerated form. The flow reversal causes a considerable boundary

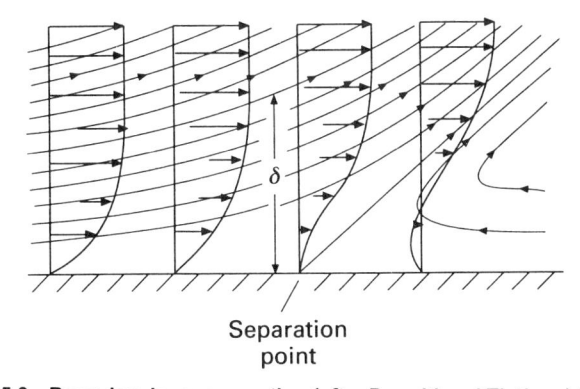

Fig. 5.9 Boundary layer separation (after Prandtl and Tietjens) (30)

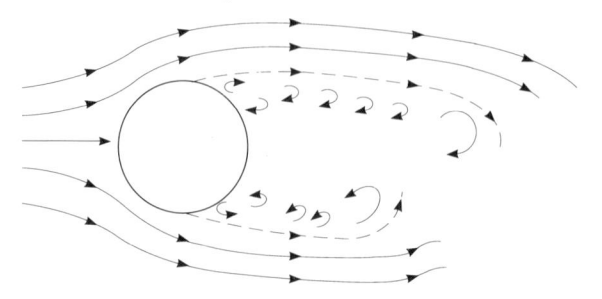

Fig. 5.10 Separating flow over a cylinder at $Re = 2000$ (after Van Dyke) (31)

layer thickening since, by continuity, the forward flow must pass over the recirculating region. An example of this is the separation which occurs on the trailing side of a cylinder (Fig. 5.10). Another example is the separation which can occur on an aerofoil at high angles of incidence. This effect is shown for a flat plate in Fig. 5.11 and it results in a loss of lift and increased drag. Figures 5.10 and 5.11 illustrate the rotating eddies, or vortices, which are created by the drag between the solid surface and the flow and which, when the boundary layer separates, result in a shear layer. This is a layer across which there is a large velocity change and within this layer the rotating eddies, or vorticity, which have been shed at the point of separation, allow this step change in velocity to occur. It is this mechanism which results in jets of air and allows one, for instance, to blow out a candle. It is the lack of shear layers which essentially prevents a jet being created by sucking.

The importance of the boundary layer is, therefore, that a small layer of fluid next to the boundary can have such a major effect on the flow. For

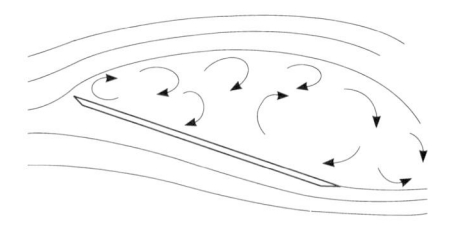

Fig. 5.11 Separation over inclined plate at 20° incidence and $Re = 10\,000$ (after Van Dyke) (31)

the engineer these effects will be important in flows over wings, fluid machinery, structures, instrumentation etc.

5.4 FLOW IN SMALL GAPS

In Appendix D the flow in a small gap is calculated. When one boundary is moving this is known as Couette flow and is likely to be laminar. The volumetric flow can be shown as

$$q_v = \frac{V_{xo}t}{2} + \frac{\Delta p t^3}{12\mu L} \text{ per unit depth of gap}$$

where L is the length of the small gap, t is its width, V_{xo} is the speed of the moving plate and Δp is the pressure difference across the passage.

This gives a simple estimate of the leakage which will occur, for instance in the gap between a moving and a stationary member in a bearing, seal or piston.

Example

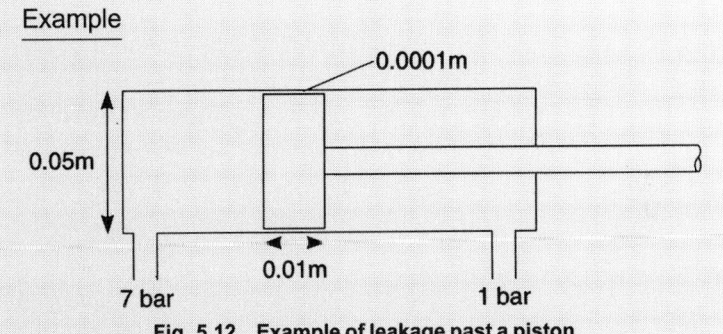

Fig. 5.12 Example of leakage past a piston

Consider a simple hydraulic piston at rest with a value of Δp = 6 bar, a clearance of 0.0001m, a piston thickness of 0.01m and diameter 0.05 m, and oil of viscosity 0.01 Pas. Hence

$q = 0.000079 \text{ m}^3 \text{ s}^{-1}$

We have taken a fluid (oil) for which density change is negligible. For air this calculation would clearly be more complex.

Many aspects of the subject of tribology concern flows in small gaps and the reader is referred to the companion volume by Summers-Smith **(32)** for a thorough introduction.

CHAPTER 6

Compressible Flow

6.1 INTRODUCTION

When a gas flows at velocities comparable with the speed of sound its behaviour is markedly different from the behaviour of incompressible fluids.

In this chapter some of these differences for special conditions are identified:

- flow through convergent nozzles;
- flow through convergent–divergent nozzles;
- flow across shock waves;
- flow in constant area ducts with friction;
- flow in constant area ducts with heat transfer.

Although idealized, the results will be of value in understanding the behaviour in pipes and valves and fluid machinery.
(Detailed derivations will be found in Appendix E)

Although high velocity flows are less common in industry, they are, nevertheless, present in some applications. This is most obviously where gases are moved in pipes at high pressures. In these cases it is possible for substantial pressure drops within the flow to occur, and with them large velocities. These occur in high-speed compressors, but also in valves, and long lengths of pipework may have significant effects in high-pressure gas flows. The critical nozzle flowmeter also depends on compressible effects.

In this chapter it is assumed that the flow at any cross-section of the duct is constant across the duct and flows parallel with the duct axis. Although not strictly true the approximation gives a good indication of actual flow changes.

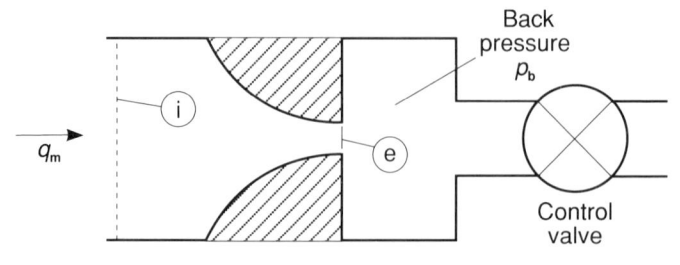

Fig. 6.1 Convergent nozzle

6.2 FLOW IN A CONVERGENT NOZZLE

To appreciate the special features of compressible flow of a gas, we start by imagining the nozzle in Fig. 6.1 set up in an experimental rig with a compressor sucking air downstream of the control valve.

Air is sucked into a nozzle which converges down to e. The pressure in the chamber downstream of e is controlled by means of a valve in the vacuum line. We start to open the valve and at the same time we plot q_m, the mass flow rate, and p_e the exit pressure, against p_b, the back pressure in the chamber before the valve (Fig. 6.2).

As p_b decreases, q_m increases and p_e has the same value as p_b. This continues until p_b has fallen to a critical pressure p_{crit}. At this point there is an abrupt change. It is found that further lowering of p_b makes no further change to q_m or p_e.

> **Notes on flow through a convergent nozzle**
> At p_{crit} – q_m reaches a fixed value and changes no more
>   – p_e ceases to be equal to p_b
> ∴ No variation in conditions downstream of the nozzle can affect q_m or the upstream properties

These are the important observations from this test. It appears that the normal information transfer from the downstream parts of the flow circuit to upstream of e have ceased and the flow upstream of e is unaffected by downstream changes.

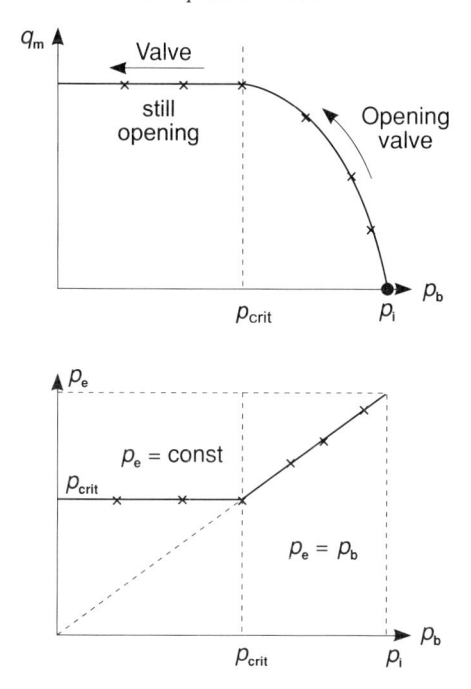

Fig. 6.2 Mass flow and exit pressure change with back pressure for convergent nozzle (after Shapiro) (33)

6.3 FLOW THROUGH A CONVERGENT–DIVERGENT NOZZLE

We now look at a second experiment. This time we consider a convergent–divergent nozzle (Fig. 6.3) which reduces in area from the inlet, i, to a minimum at the throat, t, and then increases in area to the exit, e, where it opens into the back pressure chamber.

Again the behaviour of q_m and p_e is plotted against p_b (Fig. 6.4). As before, q_m increases at first, but this time it becomes constant at a higher value of p_b and yet p_e continues to fall for some time before reaching a value at which it becomes constant.

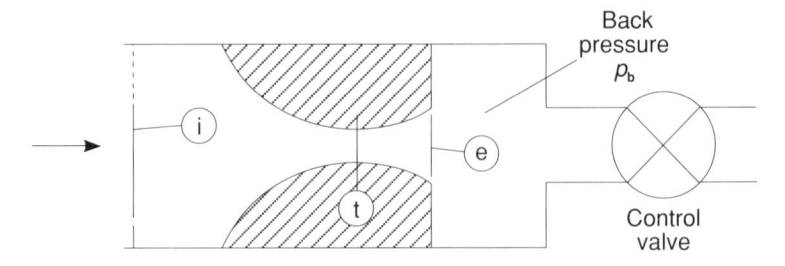

Fig. 6.3 Convergent–divergent nozzle

Notes on flow through a convergent–divergent nozzle
(i) p_b has not fallen to p_{crit} when q_m becomes constant
(ii) p_e falls below p_{crit}
(iii) increasing flow noise is followed by intermittent sound and then by silence.

Note (iii) is one of the most interesting points. If we could have done this experiment we would have heard some important changes in the noise.

(a) As the flow increased there would have been a steadily increasing noise emitted by the nozzle. The noise level would have become so loud as to preclude conversation.

(b) At about the point at which q_m became constant we would have observed that the noise changed, from the roar previously, to a loud crackling.

(c) As this point was passed the noise would have disappeared and the subsequent silence would have allowed us to hear the mechanical noise from the compressor. Without experiencing this sudden silence it is difficult to appreciate the change. However, it is so marked that the author has seen students put their hand in front of the nozzle intake to check whether air was still going in – with disastrous consequences for the manometer fluids!

We can also plot q_m against p_t and p_t against p_b (Fig. 6.5). We find that the $p_t - q_m$ curve stops at p_{crit} and that the value of p_t is less than p_b until $p_t = p_{crit}$ after which p_b drops beyond p_t.

Compressible Flow

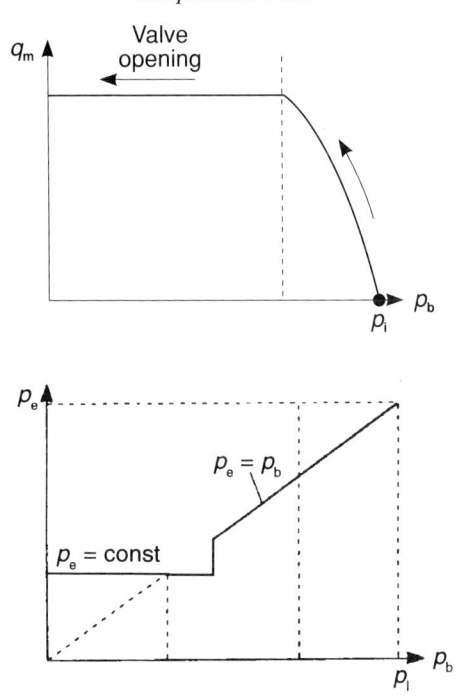

Fig. 6.4 Mass flow and exit pressure change with back pressure for convergent–divergent nozzle (after Shapiro) (33)

What is happening? As observed above, the normal information transfer mechanisms between the changes downstream and the upstream flow appear to have ceased. The loss of sound is clearly linked to this loss of information transfer, and it is intuitively reasonable to link the information transfer to the very small pressure waves which constitute sound, and to suggest that when the flow is as fast or faster than sound speed it is impossible for the waves to move upstream, and so impossible also to communicate any flow change upstream.

The velocity of sound in air at 0°C is 331 m/s, while the rms velocity of air particles at 0°C is 485 m/s or nearly 50% higher. Thus on a molecular scale, the molecular messengers, also being carried by the bulk flow, are at a speed greater than the sound waves, but moving in random directions.

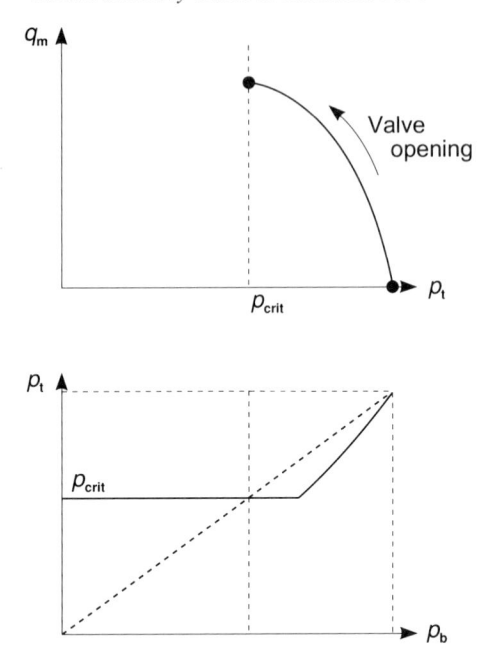

Fig. 6.5 Change of mass flow with throat pressure and throat pressure with back pressure for convergent–divergent nozzle

Presumably the sound wave causes a preferred drift of these molecules so that the velocities will not be entirely random and will result in a mean drift less than 485 m/s and of the order of the velocity of the sound wave.

6.4 EQUATIONS GOVERNING COMPRESSIBLE FLOW

The behaviour of a very important flow rig has now been reviewed. Next the equations which predict the flow are studied in an attempt to interpret what has been observed. A very useful set of equations may be derived from the following equations:

(1) continuity equation as in equation (3.13);
(2) steady flow energy equation as in equation (3.19);

(3) ideal gas equation, $pv = RT$, and equation (3.18) for a perfect gas;
(4) entropy relationship, equation (3.24),

and an expression for the Mach number, M, the ratio of flow velocity to sound speed, which is derivable by introducing the momentum across a sound wave. For the derivation of these equations the reader is referred to Appendix E, or a standard text such as that by Shapiro (**33**).

$$\frac{T_0}{T} = 1 + \frac{\gamma - 1}{2} M^2 \tag{6.1}$$

$$\frac{p_0}{p} = \left(1 + \frac{\gamma - 1}{2} M^2\right)^{\gamma/(\gamma-1)} \tag{6.2}$$

$$\frac{V}{\sqrt{(c_p T_0)}} = \sqrt{(\gamma - 1)} M \left(1 + \frac{\gamma - 1}{2} M^2\right)^{-\frac{1}{2}} \tag{6.3}$$

$$\frac{q_m \sqrt{(c_p T_0)}}{A p_0} = \frac{\gamma}{\sqrt{(\gamma - 1)}} M \left(1 + \frac{\gamma - 1}{2} M^2\right)^{-\frac{1}{2}(\gamma+1)/(\gamma-1)} \tag{6.4}$$

$$\frac{\frac{1}{2}\rho V^2}{p_0} = \frac{1}{2}\gamma M^2 \left(1 + \frac{\gamma - 1}{2} M^2\right)^{-\gamma/(\gamma-1)} \tag{6.5}$$

These five important relationships are tabulated in the gas flow tables in Appendix A. It should be noted that $\gamma = 1.4$, in Appendix A, should be used for air. The tables give the value of T, p, V, q_m/A and ρ for a given γ and M where

		UNITS
A	= area	m^2
c	= local sound speed = $\sqrt{(\gamma RT)}$	m/s
c_v	= specific heat at constant volume	J/kgK
c_p	= specific heat at constant pressure	J/kgK
M	= Mach number V/c	dimensionless
p	= pressure	Pa
p_0	= stagnation pressure	Pa

	UNITS
q_m = mass flow rate	kg/s
R = gas constant for a particular gas = R/M	J/kgK
T = temperature	K
T_0 = stagnation temperature	K
V = gas velocity	m/s
γ = ratio of specific heats c_p/c_v	dimensionless
ρ = density	kg/m^3

It should be noted that in an isentropic flow without heat transfer the stagnation temperature and pressure do not change and represent the values if the gas were to be brought to rest. However, stagnation pressure will change where irreversibility occurs, for instance through a shock wave or in a flow with friction, and stagnation temperature will change where heat transfer takes place.

Figure 6.6 shows the behaviour of M, ρ/ρ_0, and $q_m\sqrt{(c_p T_0)}/Ap_0$ when plotted against the local pressure non-dimensionalized with p_0 the inlet, or stagnation, pressure. A particularly important curve is that for $q_m\sqrt{(c_p T_0)}/Ap_0$. If at the inlet of a duct the flow is stationary then $p/p_0 = 1$. As the area decreases the pressure falls until a maximum value of the curve

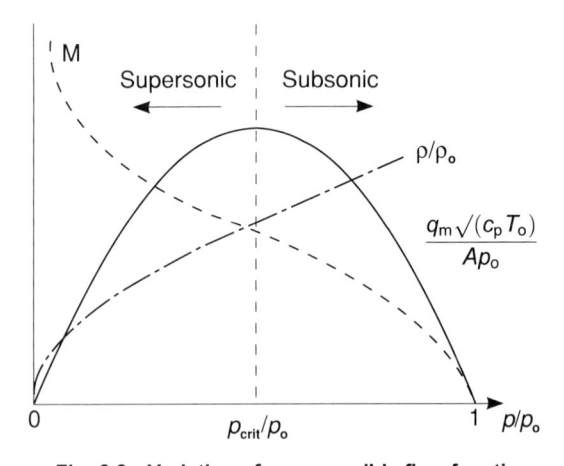

Fig. 6.6 Variation of compressible flow functions

Compressible Flow

is reached at p_{crit}/p_0 when $M = 1$ and the velocity of the gas equals the velocity of sound.

For given values of q_m, T_0, and p_0 this is the minimum value of A through which the flow can pass. Moreover, any increase in A may result in either an increase or a decrease in pressure leading to subsonic or supersonic flows. If the flow remains isentropic then this curve gives the relationship between q_m and p/p_0. When maximum q_m is reached for a given value of throat area, A, the flow is said to be choked. Equation (6.4) with $M = 1$ then forms the basis for the critical nozzle flowmeter.

6.5 EQUATIONS APPLIED TO NOZZLE FLOWS

Again the flow through the convergent nozzle (Fig. 6.7) is considered. The diagram below the outline of the nozzle shows the pressure distribution with distance through the nozzle for decreasing values of back pressure. The pressure ratios at each point through the nozzle can be used with the mass flow rate/area graph, and it can be seen that as back pressure decreases this curve is followed until the maximum point is reached. In each case the flow at exit becomes irreversible due to the sudden expansion and so ceases to follow the curves.

Figure 6.8 gives the equivalent plot for the convergent–divergent nozzle. Now it can be seen that for curves, a, b, and c, arrows show that the right limb of the mass flow rate/area curve is traversed in both directions as the area first decreases and then increases. Only for d, where the locus passes through the peak and on to the left limb, is there no return path possible since the nozzle does not again decrease in area.

It is evident that the equations predict a critical condition when the flow reaches sonic conditions at the throat, or narrowest part of the duct, and tallies with the experimental observations described earlier.

6.6 EXAMPLES OF CALCULATIONS OF NOZZLE FLOWS

The following examples indicate how the tables in Appendix A may be used:

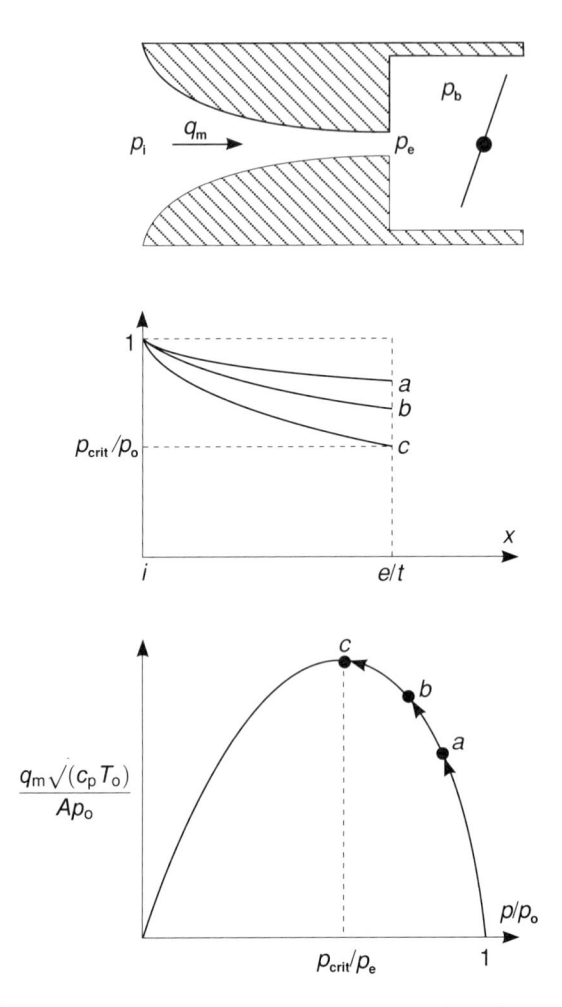

Fig. 6.7 p/p_o against distance through convergent nozzle and mass flow function against p/p_o

Compressible Flow

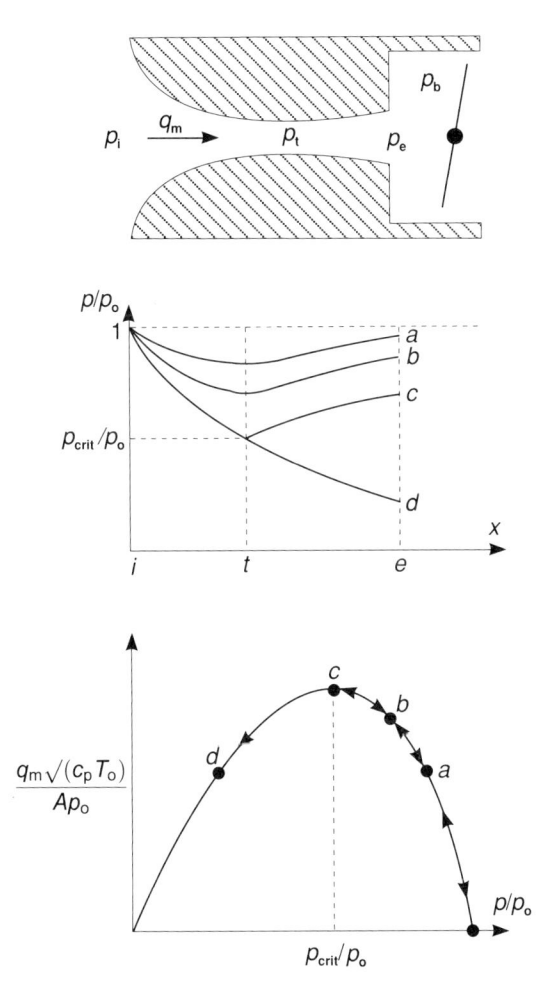

Fig. 6.8 p/p_o against distance through convergent–divergent nozzle and mass flow function against p/p_o

Example 1

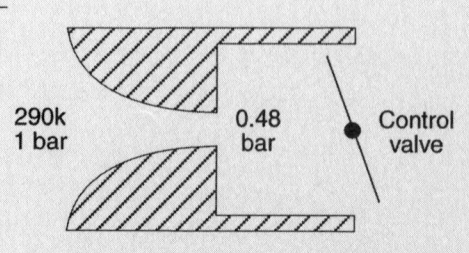

Fig. 6.9 Convergent nozzle example

A convergent nozzle with air flow (Fig. 6.9) has inlet conditions of 290K and 1 bar and a pressure in the exit chamber of 0.48 bar. The radius of the exit nozzle is 2.5 mm. What is the mass flow through the nozzle?

The first step is to check whether the nozzle is choked. We know that for a convergent nozzle the maximum pressure ratio is for $M = 1$ at exit and so from Appendix A for $\gamma = 1.4$ and $M = 1$, $p/p_0 = 0.528$. The nozzle is, therefore, choked since the actual pressure ratio is 0.48 and so the pressure in the exit plane cannot fall to the value of the pressure in the exit chamber but only to 0.528. Since the exit Mach number, $M_e = 1$

$$q_m \sqrt{(c_p T_0)}/A p_0 = 1.281$$

and

$$A = \pi (0.0025)^2$$
$$= 1.963 \times 10^{-5} \text{ m}^2$$

Inserting values we obtain

$$q_m = \frac{1.281 \times 10^5 \times 1.963 \times 10^{-5}}{\sqrt{(1006 \times 290)}} = 4.66 \times 10^{-3} \text{kg s}^{-1}$$

Example 2

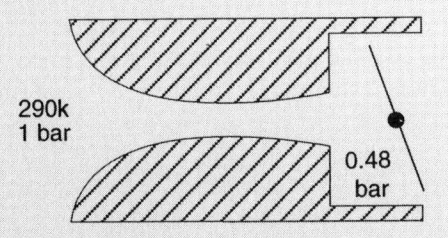

Fig. 6.10 Convergent–divergent nozzle example

A convergent–divergent duct with air flow again has inlet conditions of 290 K and 1 bar and a pressure in the exit chamber of 0.48 bar (Fig. 6.10). The exit area is 1.963×10^{-5} m^2 (2.5 mm radius) as for Example 1, but this is not any longer the throat or narrowest point in the duct. What is exit temperature and the mass flow through the duct? What is the throat area?

Since this duct is convergent–divergent the exit Mach number is not limited to $M = 1$ and so we can use the actual pressure ratio to obtain the exit Mach number from Appendix A as $M = 1.08$. From this we can obtain $T/T_0 = 0.811$ and hence the exit temperature as 235 K. We can again use the value of the mass flow function:

$$q_m\sqrt{(C_p/T_0)}/Ap_0 = 1.274 \text{ for } M = 1.08$$

and hence

$$q_m = \frac{1.274 \times 10^5 \times 1.963 \times 10^{-5}}{\sqrt{(1006 \times 290)}} = 4.63 \times 10^{-3} \text{ kg s}^{-1}$$

This is less than in Example 1 since the throat must have an area smaller than the exit area. It can be obtained from the ratio of mass flow function between throat and exit

$$A_t = \frac{1.274}{1.281} \times 1.963 \times 10^{-5} = 1.952 \times 10^{-5} \text{ m}^2$$

Between curves *c* and *d* in Fig. 6.8 there is a region which is not satisfied by the equations. The reason for this is that the equations assume that the flow has no losses: that it is adiabatic, isentropic, and reversible in thermodynamic terms. The only way that the region between c and d can be reached is by introducing the idea of a shock wave which is illustrated in Fig. 6.11.

6.7 SHOCK WAVES

A shock wave is an example of an unrestrained expansion where pressure increases within a distance of the order of a molecular mean free path and which is, therefore, irreversible and results in an entropy increase.

In order to understand the behaviour of the nozzle when isentropic conditions are not achievable, Fig. 6.11 shows how the flow develops from c in Fig. 6.8 to d as a shock wave forms and moves out of the divergent

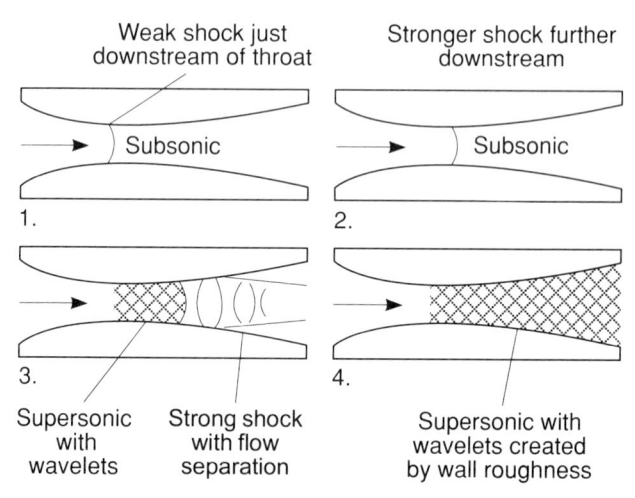

Fig. 6.11 Shock formation and movement with decreasing back pressure (after Shapiro) (33). Note that in each case flow is subsonic before the throat and supersonic immediately after the throat and before the shock wave

part of the nozzle. When the flow reaches condition c in Fig. 6.8, the flow is subsonic up to the throat, sonic at the throat, and subsonic downstream of the throat. Diagram 1 in Fig. 6.11 shows the situation when the back pressure has dropped a very small amount from p_c to p_1.

If the back pressure is now reduced slightly more (Fig. 6.12), the flow immediately downstream of the throat is supersonic for a short distance and curve d is followed further. However, the back pressure is at p_2, and in order to accommodate this outlet condition, a stronger shock forms in the duct (Fig. 6.11 Diagram 2) creating a pressure jump followed by a new subsonic curve. This subsonic curve will be disturbed by the shock upstream of it and by the diffusion. If it were isentropic, it would not follow the curve in c since the stagnation pressure, p_0, has been changed by the shock wave which is irreversible, and not isentropic.

Reducing the pressure to p_3 at outlet causes the shock to move towards the outlet of the duct and to strengthen. The shock structure is more complex, as shown in Fig. 6.11 Diagram 3. Also drawn in this diagram upstream of the shock, where the flow is supersonic, are the waves (Mach waves) which are created by slight irregularities on the walls of the duct and which move forward at the speed of sound. They therefore become more slanted as the speed of the main flow increases above the sound speed. Figure 6.12 shows the larger shock pressure jump and a new outlet curve to p_3. The flow downstream of the shock will be even more disturbed

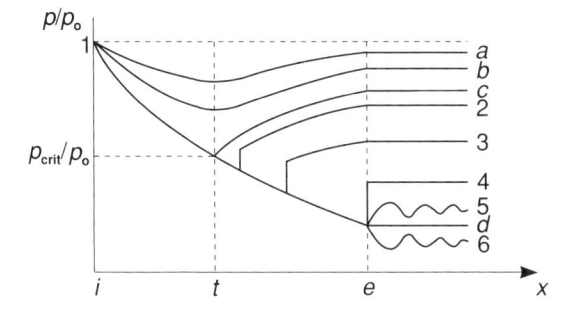

Fig. 6.12 p/p_o against distance through the nozzle when shock waves occur

in this case and the idealization of an isentropic subsonic flow to the duct outlet will be a poor one.

Finally the shock is sucked out of the duct and the whole outlet duct runs at supersonic conditions as for d. The small Mach waves are now throughout the divergent duct. Even in this condition there are some possible variations depending on whether the shock stands at the outlet, curve p_4; whether the outlet pressure is perfectly matched as for curve d; whether the pressure is between p_4 and curve d at p_5 and results in a compression wave pattern in the outlet duct; or whether the pressure is lower, p_6, and causes an expansion wave in the outlet duct.

The relationships for stationary shock waves are also tabulated in Appendix A, and equation (6.6) is derived in Appendix E.

Shock wave relations

$$M_s = \left(\frac{1 + \frac{\gamma - 1}{2} M^2}{\gamma M^2 - \frac{\gamma - 1}{2}} \right)^{\frac{1}{2}} \tag{6.6}$$

$$\frac{p_{os}}{p_0} = \left(\frac{\frac{\gamma + 1}{2} M^2}{1 + \frac{\gamma - 1}{2} M^2} \right)^{\gamma/(\gamma - 1)} \left(\frac{2\gamma}{\gamma + 1} M^2 - \frac{\gamma - 1}{\gamma + 1} \right)^{1(1-\gamma)} \tag{6.7}$$

$$\frac{p_s}{p} = \frac{2\gamma}{\gamma + 1} M^2 - \frac{\gamma - 1}{\gamma + 1} \tag{6.8}$$

$$\frac{p_{os}}{p} = \left(\frac{\gamma + 1}{2} M^2 \right)^{\gamma/(\gamma - 1)} \left(\frac{2\gamma}{\gamma + 1} M^2 - \frac{\gamma - 1}{\gamma + 1} \right)^{1(1-\gamma)} \tag{6.9}$$

$$\frac{T_s}{T} = \frac{\gamma - 1}{(\gamma + 1)^2} \frac{2}{M^2} \left(1 + \frac{\gamma - 1}{2} M^2 \right) \left(\frac{2\gamma}{\gamma - 1} M^2 - 1 \right) \tag{6.10}$$

Example

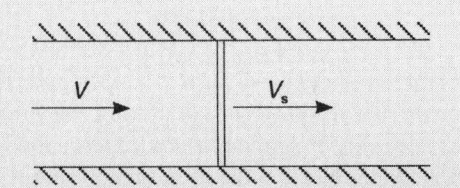

Fig. 6.13 Shock wave held stationary in a duct

A shock wave moves through air at Mach 2. What is the pressure and temperature after the shock? What is the velocity of the air after the shock? Ambient conditions are 1 bar and 293K.

In the tables in Appendix A stationary shocks are considered, as in Fig. 6.13, and for a shock of $M = 2$, $p_s/p = 4.50$, $T_s/T = 1.688$. p and T are the static values of pressure and temperature and so the values downstream of the shock will be $p_s = 4.5$ bar and $T_s = 495$K. The passage of the shock will also result in a flow in the same direction as the shock due to the requirement to draw in air to fill the higher density after the shock. This can be calculated from the velocities of flow into and out of a stationary shock. These are then transformed into those for a moving shock. Since $T/T_0 = 0.556$ and for a stationary shock $T_0 = 293/0.556 = 527$K

at inlet $\dfrac{V}{\sqrt{(c_p T_0)}} = 0.943$ and $V = 687$ m/s

Since $M_s = 0.577$ at outlet $\dfrac{V_s}{\sqrt{(c_p T_0)}} = 0.353$

and $V_s = 257$ m/s

Thus, by assuming the air is stationary ahead of the shock, the shock will move at 687 m/s through still air, and the velocity of the air after the passage of the shock will be $687 - 257 = 430$ m/s.

Such an example can be conceived as a result of an explosion where the front is part of an increasing spherical shock wave.

6.8 EXAMPLE OF A CONVERGENT–DIVERGENT DUCT CALCULATION WITH SHOCK WAVE

The calculations of convergent–divergent ducts and shock waves are brought together in this section by calculating the flow in such a duct with a shock wave. In this example a knowledge of the Mach number at inlet to the shock and in the exit plane is assumed. This allows the solution to be achieved without iteration. The duct is shown in Fig. 6.14(a) which also labels the stations and the labels will be used as subscripts.

First of all the flow in the duct upstream of the shock is solved. Since the flow between the throat and the shock is supersonic and assuming that the inlet flow is subsonic, the Mach number at the throat must be unity. So from the tables (Appendix A) we obtain

for $M = 1$ $(p_t/p_0) = 0.528$

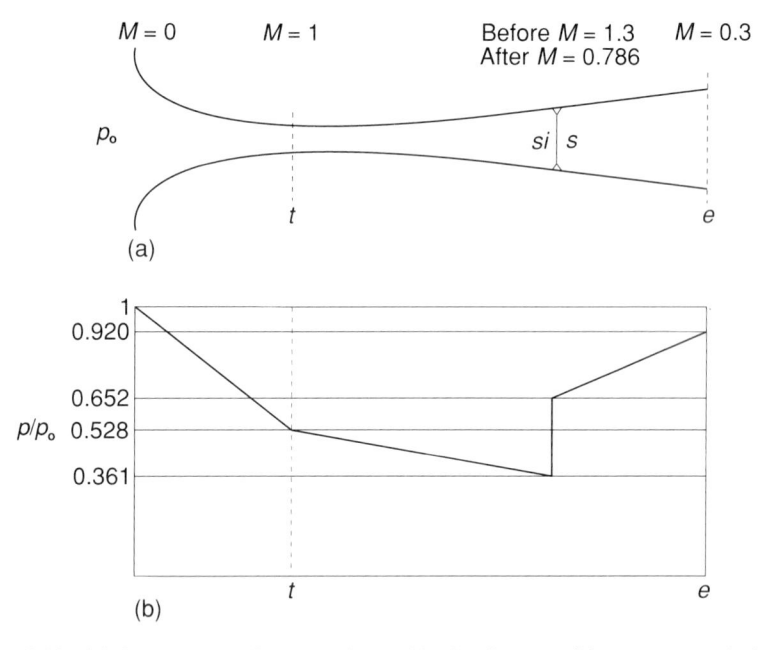

Fig. 6.14 (a) Convergent–divergent duct with shock wave; (b) pressure variation through duct

Compressible Flow

and we can obtain the pressure at shock inlet

for $M = 1.3$ $(p_{si}/p_0) = 0.361$

We can now use the table for conditions across the shock to obtain:

for $M = 1.3$ $M_s = 0.786$ and $(p_s/p_{si}) = 1.805$

From which we obtain

$p_s/p_0 = p_s/p_{si} \times p_{si}/p_0 = 1.805 \times 0.361 = 0.652$

For the flow immediately downstream of the shock we obtain from the normal isentropic table

for $M_s = 0.786$ $p_s/p_{0s} = 0.665$

For outlet we have

$M = 0.3$ so $p_e/p_{0s} = 0.939$

We require p_e/p_0

$= p_e/p_{0s} \times p_{0s}/p_s \times p_s/p_{si} \times p_{si}/p_0$

$= 0.939 \times \dfrac{1}{0.665} \times 1.805 \times 0.361$

$= 0.920$

The graph of p/p_0 can now be plotted against distance through the nozzle, as shown in Fig. 6.14(b).

We can also obtain the area change through the duct by using the table for $q_{mi} \dfrac{\sqrt{(c_p T_0)}}{A_t p_0}$

For the throat

$q_{mi} \dfrac{\sqrt{(c_p T_0)}}{A_t p_0} = 1.281$

For the shock inlet where $M = 1.3$

$q_{mi} \dfrac{\sqrt{(c_p T_0)}}{A_{si} p_0} = 1.201$

Therefore

$\dfrac{A_{si}}{A_t} = \dfrac{1.281}{1.201} = 1.067$

Across the shock using the above figures we can obtain

$$p_{0s}/p_0 = p_{0s}/p_s \times p_s/p_{si} \times p_{si}/p_0 = 0.980$$

After the shock

for $M = 0.786$ $\quad q_{mi}\dfrac{\sqrt{(c_p T_0)}}{A_s p_{0s}} = 1.226$

which also equals

$$q_m \dfrac{\sqrt{(c_p T_0)}}{A_{si} p_0} \times \dfrac{p_0}{p_{0s}} = 1.201/0.980 = 1.226$$

And at outlet for $M = 0.3$

$$q_m \dfrac{\sqrt{(c_p T_0)}}{A_e p_{0s}} = 0.629$$

Thus

$$\dfrac{A_e}{A_s} = \dfrac{1.226}{0.629} = 1.949$$

And hence, since $A_s = A_{si}$

$$\dfrac{A_e}{A_t} = 1.949 \times 1.067 = 2.080$$

We have therefore obtained the pressure variation through the duct and the area related to the throat area.

Had we known the exit pressure and area and that a shock was standing in the divergent part of the duct, the following procedure would have been followed:

(1) assume a position of the shock either in terms of M or A;
(2) obtain conditions following the shock;
(3) knowing the exit area of the duct obtain the pressure, p_{e1};
(4) recalculate from, (1)–(3) to obtain a new exit pressure, p_{e2}, for a new position of the shock;
(5) plot the two values of pressure, p_{e1} and p_{e2} as in Fig. 6.15 and obtain a better approximation using the given value of pressure, p_e, to obtain a better estimate of the shock position;
(6) continue until the result is of adequate accuracy.

Compressible Flow

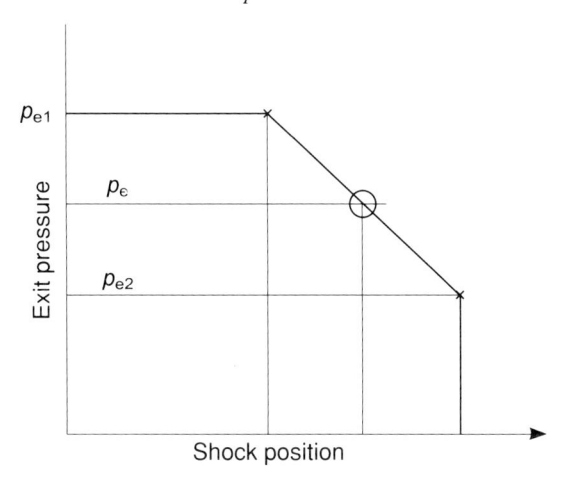

Fig. 6.15 Iterative approach to obtain the position of the shock wave from the outlet conditions for a convergent–divergent duct

6.9 CONSTANT AREA DUCTS

Before leaving compressible flow, two further flows need to be considered. These occur in a constant area duct. Figure 6.16 shows the two cases of interest, the first with friction and no heat transfer, the second with heat transfer and no friction. The situation where gas is flowing through a long

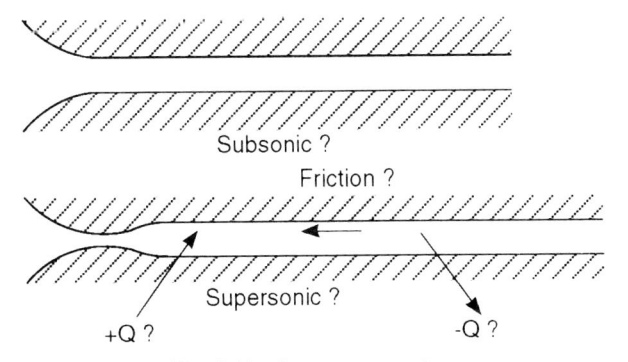

Fig. 6.16 Constant area ducts

pipe with friction occurs in many plants and it is worth knowing the possible changes which may occur in the Mach number and velocity. The same must be true for heating and cooling a gas as it flows through a duct. A duct with a contraction at the inlet will result in an initial velocity which is, at most, sonic. A duct which has a convergent–divergent nozzle at inlet may have supersonic conditions in part of it.

6.10 CONSTANT AREA DUCT WITH FRICTION–FANNO LINE

If gas flows through a constant area duct with friction, and if it is subsonic at inlet (Fig. 6.17), its Mach number increases (curve a) until the highest Mach number that can be attained is unity, or sonic conditions, at the outlet (curve b). If the inlet flow is supersonic (Fig. 6.18), then the Mach number will decrease along the duct until sonic conditions are attained at the outlet (curve c, for the duct ending at c_2). If the length of the duct is too great, then the flow passes through a shock wave and the flow becomes subsonic attaining sonic conditions at the outlet (curve d, for the duct

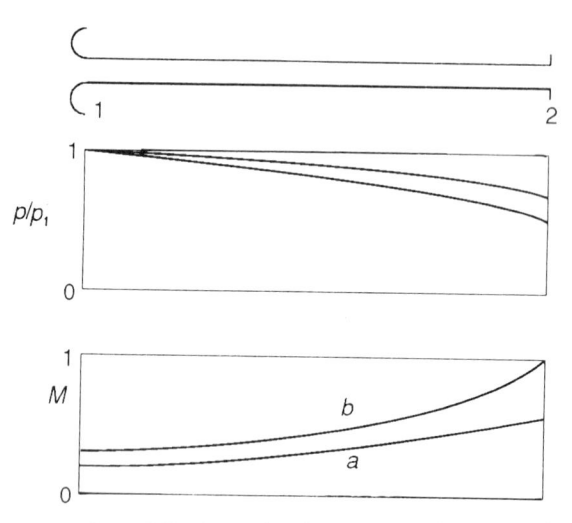

Fig. 6.17 Pressure ratio and Mach number for subsonic flows with friction (after Shapiro) (33)

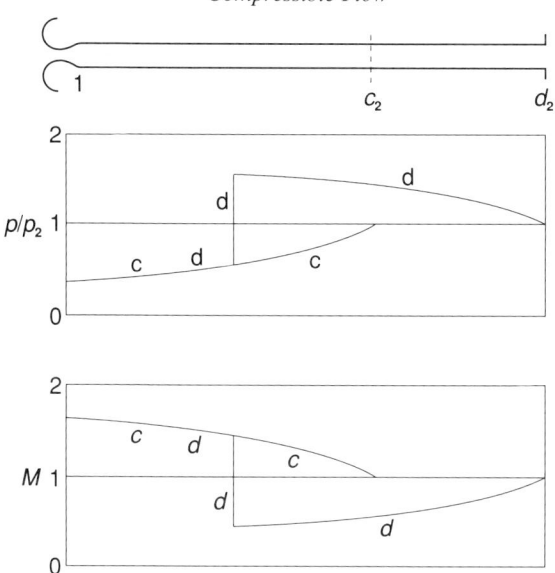

Fig. 6.18 Pressure ratio and Mach number for supersonic flows with friction (after Shapiro) (33)

ending at d_2). This is shown in Fig. 6.18. In practice, supersonic flow tends to be short-lived, probably due to boundary layer disturbance (**13**).

It should be recalled that this is *not* an isentropic process and that entropy of the gas is increasing as the gas passes down the duct. In addition the Mach number tends to unity. Figure 6.19 shows curves of temperature against entropy in the duct. Since the flow is adiabatic and irreversible the process must always move to higher values of entropy, s. If the flow is subsonic initially (curves a and b) then the maximum flow is that when the Mach number at exit is unity and the flow is choked, b_2. If the flow is supersonic at entry then more options are possible. The flow can be supersonic throughout with Mach number of one at exit (curve c). Alternatively a shock occurs and the subsequent subsonic flow again results in Mach one at exit (curve d starting at d_1 and ending at d_2).

From the gas flow tables (Appendix A) we can now use an expression which relates the friction coefficient to the maximum length of duct to achieve sonic flow at outlet (equation (6.10)).

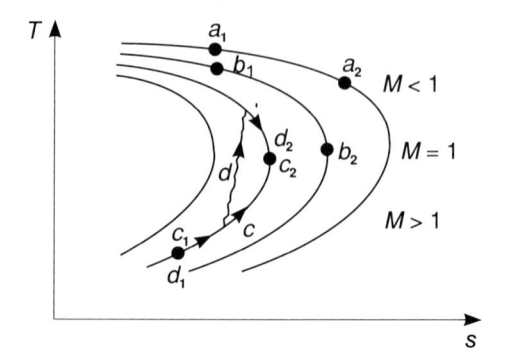

Fig. 6.19 Temperature against entropy for flow with friction (after Shapiro) (33)

Equation for flows in constant area ducts with friction

$$\frac{4fL_{max}}{D} = \frac{1-M^2}{\gamma M^2} + \frac{\gamma+1}{2\gamma} \ln \frac{(\gamma+1)M^2}{2\left(1+\frac{\gamma-1}{2}M^2\right)} \qquad (6.10)$$

Note: the friction coefficient $f = f_D/4$

In this expression L_{max} is the maximum length of the duct in order that sonic conditions can be attained at the exit without shock waves. In many cases we shall not have sonic conditions at outlet, so we have to use the tables in a rather subtle way which will be explained by an example.

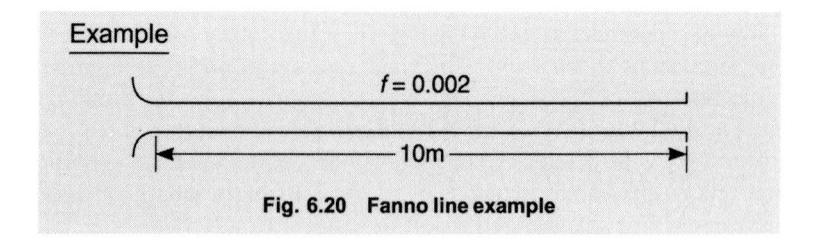

Example

$f = 0.002$

10m

Fig. 6.20 Fanno line example

After a contraction the duct is as in Fig. 6.20. The inlet stagnation conditions are $p_0 = 1$ bar and $T_0 = 293$K. The mass flow rate is 5.69×10^{-2} kg/s, the duct diameter is 0.025 m, and the area is, therefore, 4.91×10^{-4} m². The duct length is 10 m and the value of $4f = 0.008$. Find the value of Mach number at duct exit.

To obtain the Mach number at the entry to the constant area tube, we use the mass flow function $q_m \sqrt{(c_p T_0)}/A p_0$ with the area of the duct to obtain the Mach number at exit from the isentropic inlet contraction flow

$q_m \sqrt{(c_p T_0)}/A p_0 = 0.629$

∴ $M = 0.3$

We now turn to the function $4fL_{max}/D$ for $M = 0.3$ and obtain the value from the table

$4fL_{max}/D = 5.299$

from which we obtain $L_{max} = 16.56$ m. Now, this is the subtle point about the use of these tables. We require the outlet Mach number for a duct of length 10 m. Were the tube 16.56 m long we should have the answer $M = 1$! To find the value of Mach number at exit from the 10 m duct we note that this will be the same as the Mach number at inlet to a duct with $L_{max} = 6.56$ m. For this length of duct we calculate

$4fL_{max}/D = 2.099$

and from the tables with this value we obtain the Mach number we require

$M = 0.41$

Note that for a non-circular duct $D = 4 \times$ Area/Perimeter. Remember, as noted in Chapter 4, that the value of friction coefficient varies with different authors. The value in this chapter is that used in the gas tables in Appendix A. The value used by Miller (**13**) for losses in pipes is four times this value and in this book has been given the subscript D.

An example giving the outlet pressure would have taken longer to solve, since an iterative method would probably be necessary. Flows which start supersonic are more complicated to solve but can still be done by iteration. The shock position needs to be guessed and patched into the supersonic and subsonic curves.

6.11 CONSTANT AREA DUCT WITH HEAT TRANSFER – RAYLEIGH LINE

If a gas flows through a constant area frictionless duct with heating the stagnation temperature, the velocity and the Mach number all increase for subsonic flow. However, while the stagnation temperature increases in supersonic flow, the velocity and the Mach number decrease. When cooling takes place the reverse happens. Figure 6.21 shows this and Table 6.1 sets out the changes.

Figure 6.21 shows how T increases for $M < 1$ to the left of the curve's maximum point and decreases when to the right for heating; and the opposite for cooling – an intuitively strange result. Note also that at supersonic flows cooling causes the Mach number to increase!

The example below shows how to use the table. In this case equation (6.11) is used which is also tabulated in Appendix A.

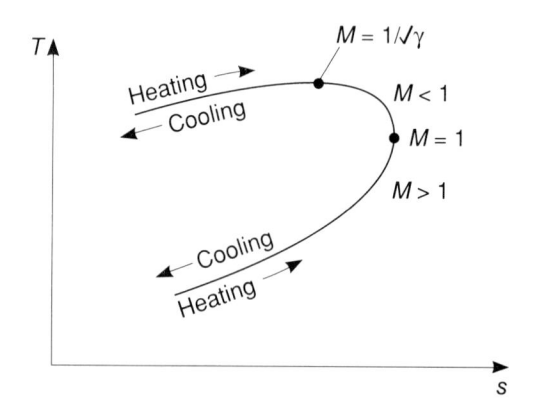

Fig. 6.21 Temperature against entropy for one set of conditions of the duct for frictionless flow with heat transfer – Rayleigh line (after Shapiro) (33)

Compressible Flow

Table 6.1 Changes in a constant area duct without friction due to heating and cooling (after Shapiro) (33)

	Heating $M < 1$	Heating $M > 1$	Cooling $M < 1$	Cooling $M > 1$
T_0	Increases	Increases	Decreases	Decreases
M	Increases	Decreases	Decreases	Increases
T	*	Increases	**	Decreases
p	Decreases	Increases	Increases	Decreases
p_0	Decreases	Decreases	Increases	Increases
V	Increases	Decreases	Decreases	Increases

* Increases for $M < 1/\sqrt{\gamma}$; decreases for $M > 1/\sqrt{\gamma}$
** Decreases for $M < 1/\sqrt{\gamma}$; increases for $M > 1/\sqrt{\gamma}$

Equation for flows in constant area ducts with heat transfer

$$\frac{F}{q_m\sqrt{(c_p T_0)}} = \frac{\sqrt{(\gamma-1)}}{\gamma}\frac{1+\gamma M^2}{M}\left(1+\frac{\gamma-1}{2}M^2\right)^{-\frac{1}{2}} \quad (6.11)$$

Note: F, the impulse function, remains constant along a duct which is frictionless

Example

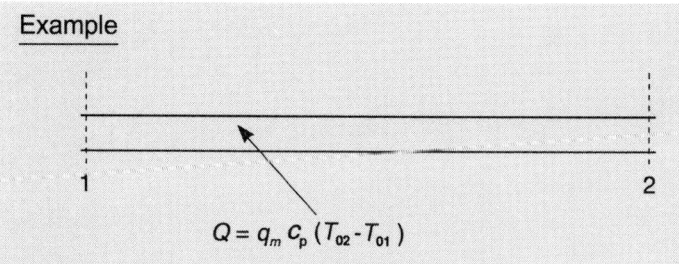

$Q = q_m c_p (T_{02} - T_{01})$

Fig. 6.22 Rayleigh line example

We refer to Fig. 6.22 which shows the duct and we wish to find M_2 if T_{01} = 293K, M_1 = 0.2, and Q = 400 kJ/kg.

We can calculate the change in stagnation temperature and hence

$$T_{02} = \frac{400}{1.006} + 293 = 691$$

The Mach number allows us to obtain the value of

$$\frac{F}{q_m \sqrt{(c_p T_{01})}}$$

from the tables in Appendix A as 2.376.

With this value and the ratio of stagnation temperature between inlet and outlet, 0.42, we can obtain the value of

$$\frac{F}{q_m \sqrt{(c_p T_{02})}} \text{ as } 2.376 \times 0.65 = 1.547$$

$$\therefore \quad M_2 = 0.33$$

CHAPTER 7
Oscillations and Waves in Fluids and Water Hammer

This chapter gives a brief overview of a huge subject. The following are considered:
- surface waves;
- pulsation problems in pipes;
- water hammer;
- fluid oscillations;
- ultrasound.

7.1 INTRODUCTION

Waves on water are one of the few examples of waves which can be observed in everyday life. Other waves and oscillations are more evident from their effects, such as pressure variations. Pressure waves in a liquid may lead to water hammer. This can set up oscillations in pipework, which sound like hammering in the pipes. Unsteady flow can sometimes set up oscillations in pressure lines which lead, for instance, to a manometer and the fluctuations can result in reading errors in the manometer. Another oscillation is that resulting in Aeolian tones – vortex shedding. This is closely related to the oscillation in the fluidic flowmeter. All these effects are complicated and the treatment here will, in the main, be descriptive.

7.2 SURFACE WAVES

Figure 7.1 shows the motion of particles in a moving wave and makes the point that motion continues but with exponentially decreasing amplitudes with increasing depth. The wave velocity, v, for a very deep

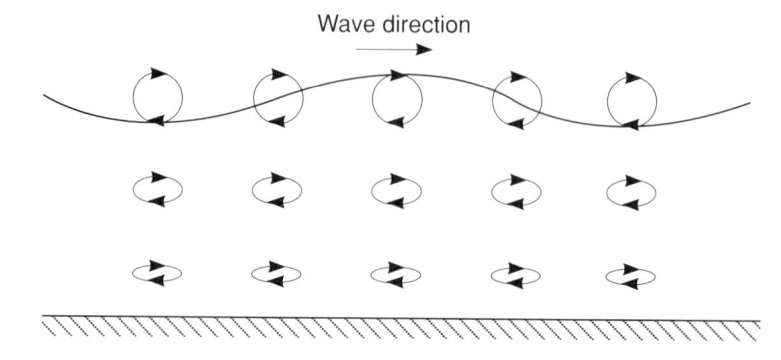

Fig. 7.1 Motion of particles in a surface wave (after Van Dyke) (31)

sea (depth/wavelength large) is

$$v^2 = \frac{g\lambda}{2\pi} \tag{7.1}$$

where g is the gravitational acceleration and λ is the wave length. For a very shallow sea

$$v = \sqrt{(gh)} \tag{7.2}$$

where h is the depth of water. An important dimensionless group, the ratio of kinetic to potential energy, is the Froude number, Fr, which is also the ratio of liquid velocity squared to wave speed squared

$$Fr = \frac{V^2}{gh} \tag{7.3}$$

It is important because changes in flow are communicated upstream by waves, and so when $Fr > 1$ the wave speed is less than the liquid speed and waves cannot propagate upstream. It, therefore, has a similar effect in free surface liquid flows to Mach number in compressible gas flows. We shall return to consideration of open channel flows in Chapter 8.

> **Example**
>
> It is found that a heavy displacement boat has a maximum speed given approximately by the speed of waves of wavelength equal to its length. Find the maximum speed of a boat of 10 m length.
>
> $V \approx \sqrt{(9.81 \times 10/2\pi)} = 3.95$ m/s ≈ 8 UK knots

7.3 PULSATION PROBLEMS IN PIPES

Care should be taken in cases where the flow is known to be unsteady in internal pipe flows, as pulsations can affect the performance of instrumentation, and standing waves may be set up in impact tubes transferring pressure measurement between line and pressure cell or manometer. In severe cases pulsation can actually destroy the pipe system. Pumps and compressors may cause pulsation but the pipe layout may increase the problem.

7.4 WATER HAMMER

The phenomenon of water hammer can be heard in domestic water systems when a valve is shut too quickly, and oscillations may, on occasion, be set up by the ball valve in a roof tank resonating with the surface waves in the tank and causing repeated hammering. (This can sometimes be solved by attaching a baffle which sits in the tank attached to but under the ball and damps the oscillations.) The forces involved can be destructive and the phenomenon should not be allowed to occur in industrial pipework. An approximate expression for the pressure change resulting from the wave created by a sudden flow rate change is

$$p - p_0 \approx \rho V c \qquad (7.4)$$

The right side of the equation is the rate of change of momentum of a column of velocity V, density ρ, and length c – the distance traversed by a pressure wave of velocity c in one second. Such a pressure wave travelling at approximately the speed of sound in liquid, c, is given by

$$c = \sqrt{\frac{\kappa}{\rho}} \tag{7.5}$$

where κ is the bulk modulus. The value of c will actually be less than this value due to pipe wall flexibility.

> ### Example
>
> Find the force on the end of a 25 mm diameter pipe if water flowing at 1 m/s is stopped suddenly. Take $\kappa = 2$ GN/m².
>
> $p - p_0 = V\sqrt{(\rho\kappa)}$
> $ = \sqrt{(1000 \times 2 \times 10^9)}$
> $ = 1.4 \times 10^6$ N/m²
> $ = 14$ bar
>
> The force on the pipe is 687 N (equal to the weight of a 70 kg load).

One way of providing damping is to incorporate a branch into the pipe system with a surge tank as shown in Fig. 7.2.

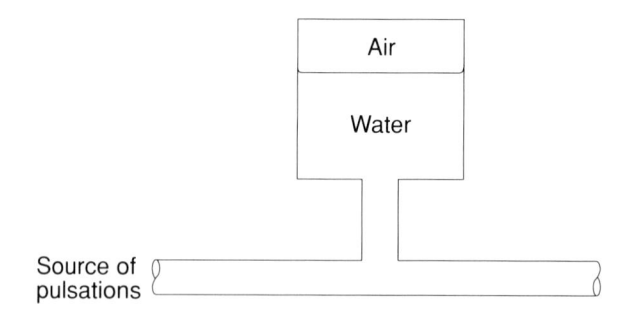

Fig. 7.2 Surge tank

7.5 FLUID OSCILLATIONS

Section 5.3 mentioned random attachment of the flow which, in a wide angle diffuser, arbitrarily forms a submerged jet on one wall or the other. By a suitable feedback path the jet can be made to jump from one wall to the other (Fig. 7.3) and the oscillation frequency is found to be closely proportional to the flow velocity. This is intuitively reasonable, since the increase in flow rate will reduce the time for flow through the feedback path and will increase frequency. This is the basis of the fluidic flowmeter.

A related effect is that of vortex shedding which occurs around objects in the path of a flow. Its most common occurrence is around cylindrical wires which results in an audible tone, for instance, from the rigging of a yacht in a high wind. Figure 5.10 shows the formation of vortices as the shear layers roll up. Figure 7.4 is an idealized diagram of this.

The vorticity shed will be of order V/δ, where δ is the shear layer thickness. The rate of shedding will be approximately the mean velocity in the shear layer, $V/2$, multiplied by the layer thickness, δ. Thus vorticity shedding rate is

$$\frac{V}{\delta} \times \frac{\delta V}{2} = \frac{V^2}{2}$$

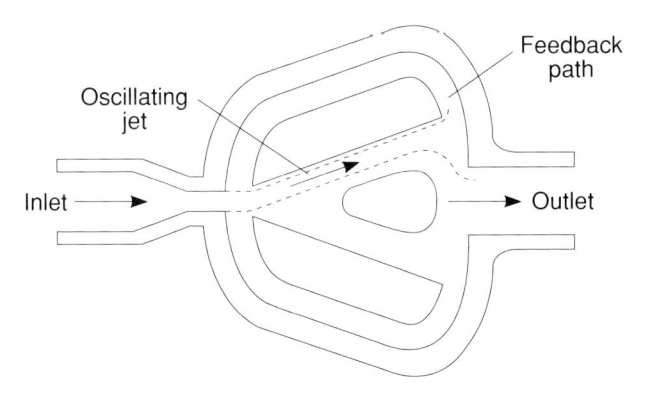

Fig. 7.3 Fluidic oscillator

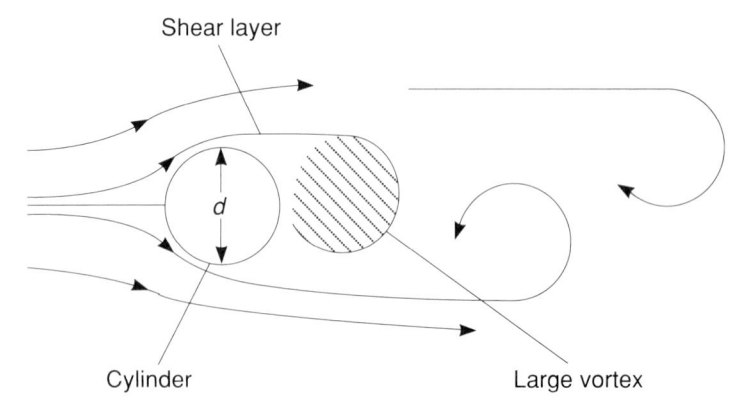

Fig. 7.4 **Vortex shedding from cylindrical bluff body**

The time for this to fill a vortex of diameter d is approximately

$$\frac{\pi d V}{V^2/2} = 2\pi d/V$$

Thus an order of magnitude value for the frequency is

$$f = \frac{V}{2\pi d}$$

The shedding frequency is given by the Strouhal number

$$S = \frac{fd}{V} \tag{7.6}$$

Our calculation gives a value of $S = 1/2\pi = 0.16$, compared with experimental value (7) for a circular cylinder at a Reynolds number in the range 300–100 000 of 0.18.

This effect can result in severe lateral forces on a structure of any cross-section and may lead to failure as in the case of the Tacoma Narrows Bridge when crosswinds caused the road to oscillate. The oscillation can sometimes be heard around tall rectangular buildings, as the flow pattern changes with each vortex shed. Chimneys are often constructed with helical spoilers around them to cause the shedding to become incoherent, and the forces, in consequence, to be reduced. Various designs of spoilers have been proposed for subsea cables to prevent damage caused by flow-induced oscillations.

7.6 ULTRASOUND

> **Uses of ultrasound in fluids:**
> - speed measurement: industrial flow, blood flow;
> - distance measurement: level, range;
> - position sensing: impedance change across an interface;
> - imaging: medical diagnostics;
> - communication: remote signalling through air;
> - cleaning: ultrasonic cleaning baths

Longitudinal waves in gases and liquids at frequencies above the audible range (which is 20 Hz to 20 kHz) are known as ultrasound and offer a very useful tool for a variety of applications. Ultrasound is usually created by piezoelectric crystals which operate at 40 kHz and upwards for gases and 100 kHz to a few MHz for liquids. Very high power ultrasound in a liquid bath creates local cavitation which cleans objects in the bath.

The ultrasonic waves will have frequency f, and a wavelength λ, so that the relationship between sound speed, frequency and wavelength is given by

$$c = \lambda f \qquad (7.7)$$

They will also have a certain amplitude which will be attenuated with transmission, an effect which can be used to obtain density or composition in a flow.

Ultrasound is used for sensing distance and position as in liquid level in a tank (see Chapter 2). Pulses of ultrasound are transmitted in the direction to be measured and the time of arrival of the returning pulse reflected from an interface is used to obtain the elapsed time and, with a knowledge of the sound speed, the distance.

It is used for velocity measurement in liquids and gases as described briefly at the end of Chapter 1 and in Baker (**1**). In velocity measurement, the pulses are carried by the fluid, so that elapsed time in the flow direction will be less than against the flow. The difference can be used to obtain the flow rate.

The beam of ultrasound will be bent as it crosses in a direction perpendicular to the flow, and the degree of bending may also be used as a measure of the flow rate by sensing the change in strength of the received signal at two sensors displaced in the flow direction.

As with other signals, ultrasound can be used with correlation techniques to obtain the velocity of a fluid by correlating two parallel beams displaced in the direction of the flow. The correlation will sense the pattern of amplitude variation in each beam and compare these to obtain the time of passage between the beams.

The doppler effect can also be used to obtain the velocity of a moving object. The reflected pulse will experience a frequency change. This is due to the doppler shift which occurs when sound is reflected from a moving object and, although the sound speed is unchanged, the time between reflection of wave peaks will be increased or decreased by the movement.

The piezoelectric crystal produces a beam of ultrasound which spreads, as with other waves, according to whether the source area is large compared with the wave length. Thus for ultrasound in air with a frequency of 20 kHz and a sound speed of 343 m/s (at 20°C), the wave length will be $343/20,000 \simeq 17$ mm. This will be of the same order of size or larger than the transducer, and the wave front will spread over a large angle with poor directionality. For water with a frequency of 1 MHz and a sound speed of 1414 m/s, the wave length will be 1.4 mm, and will be small compared with a crystal of, say, 10 mm diameter, leading to less beam spreading and a greater directionality of the beam. Lower frequencies are used in gases due to the high attenuation at high frequency.

When ultrasound crosses an interface between two different fluids or between a fluid and a solid, there is both transmission and reflection at the interface. The transmission will be greater if the density times wave speed of the two materials are similar in size. Thus it proves difficult to transmit ultrasound from air, through metal, and then into a transducer. On the other hand it is possible to transmit from liquid through a solid and then to a transducer. The characteristic of the materials which is relevant to this is the impedance.

> Impedance
> = density of material through which ultrasound is transmitted × velocity of ultrasound in the material
>
> $Z = \rho \times c$
>
> $Z_{AIR} = 343 \times 1.19 = 408 \text{ kgm}^{-2}\text{s}^{-1}$
> $Z_{WATER} = 1414 \times 1000 = 1.41 \times 10^6 \text{ kgm}^{-2}\text{s}^{-1}$
> $Z_{STEEL} = 5625 \times 8000 = 45 \times 10^6 \text{ kgm}^{-2}\text{s}^{-1}$

The proportion of ultrasound power transmitted is given by Asher (**34**)

$$P_T = \frac{4Z_1 Z_2}{(Z_1 + Z_2)^2} \tag{7.8}$$

and that reflected is

$$P_R = \frac{(Z_1 - Z_2)^2}{(Z_1 + Z_2)^2} \tag{7.9}$$

Thus while transmission from steel to water is about 12 percent, that from air to steel is negligible.

The scope of this book does not allow consideration of other ultrasonic applications.

CHAPTER 8

Open Channel Flow

A brief review of flows
- in open channels
- weirs and flumes
- a hydraulic jump

8.1 INTRODUCTION

Flows in open channels are particularly important in water treatment works, where much of the movement of the water takes place in open, or partially filled, conduits. Because the surface is subject to atmospheric pressure, the level of the surface relates to the velocity of the flow. This relationship can be obtained from the Bernoulli equation (3.27)

$$\frac{p_2 - p_1}{\rho} + \frac{(V_2^2 - V_1^2)}{2} + g(z_2 - z_1) = 0$$

At station 1, Fig. 8.1, velocity is assumed to be zero and the surface level or head is given by $z_1 = H$. If Bernoulli is then applied to the surface streamline, where $p_1 = p_2 = p_0$, the atmospheric pressure, then for

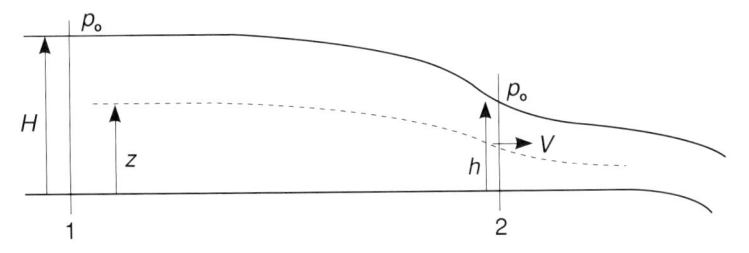

Fig. 8.1 Open channel flow

station 2

$$V = \sqrt{\{2g(H-h)\}} \tag{8.1}$$

where h is the surface level and V is the velocity at station 2. If the channel has a width b, then the volumetric flow is

$$q_v = bh\sqrt{\{2g(H-h)\}} \tag{8.2}$$

The maximum flow rate can be found by differentiating this expression

$$\frac{dq_v}{dh} = b\sqrt{(2g)}\{(H-h)^{1/2} - \tfrac{1}{2}h(H-h)^{-1/2}\}$$

and for the maximum value of the flow rate $dq_v/dh = 0$, and we obtain

$$h = h_c = 2H/3 \tag{8.3}$$

and this is called the critical depth. From this the maximum value of flow rate can be obtained as

$$q_{vmax} = bH\sqrt{(8gH/27)} \tag{8.4}$$

8.2 WEIRS AND FLUMES

Equation (8.4) is important in the measurement of flow in open channels using broad crested weirs and flumes, Fig. 8.2. If the value of H, the head above the weir or the bottom of the flume is measured, and the value of b is known, the value of q_{max} can be obtained provided that the flow measuring section is running at maximum flow rate. This is ensured by providing a sufficient fall after the flow measurement section. The measurement of H may not be possible and a modified formula using h, the head over the weir, is required. The full formula with discharge coefficients can be obtained from the standard BS3680 or from reference (35). The velocity for these conditions from equations (8.1) and (8.3) is

$$V = \sqrt{(gh_c)} \tag{8.5}$$

and from equation (7.3) it is apparent that the Froude number for this flow is unity. Flows with $h > h_c$ have a Froude number less than unity and are called tranquil, subundal or subcritical flows. Flows with $h < h_c$ have a Froude number greater than unity and are called shooting, superundal or

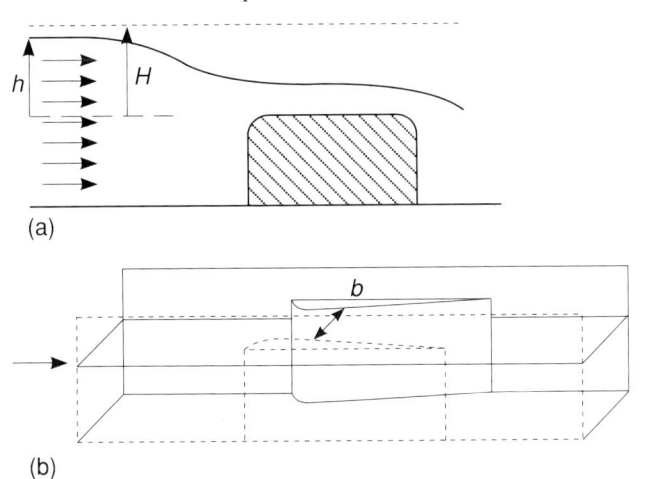

Fig. 8.2 (a) Broad crested weir, (b) flume

supercritical flows. There is a parallel with the sub- and supersonic flows in gases. In subundal flows the wave velocity is sufficient for changes downstream to be communicated upstream, whereas in superundal flows the velocity of the liquid is too great for downstream changes to be communicated upstream. The critical flume, therefore, provides a parallel with the critical nozzle. Equations for V-notch and other weirs may be obtained from reference (**35**) or from the BS and ISO standards.

8.3 HYDRAULIC JUMP

The behaviour of the flow through an hydraulic jump is obtained by using the equation for conservation of momentum. Figure 8.3 is a diagram of an hydraulic jump. Such a jump can be observed in a sink when a tap runs and the water is deflected outwards in a thin layer. A short distance radially outward, the high-speed shooting flow gives way to a wave front or jump after which the flow is deeper. In deriving the equation using conservation of momentum (since an energy loss occurs preventing the use of Bernoulli's equation), it is assumed that the friction around the channel

An Introductory Guide to Industrial Flow

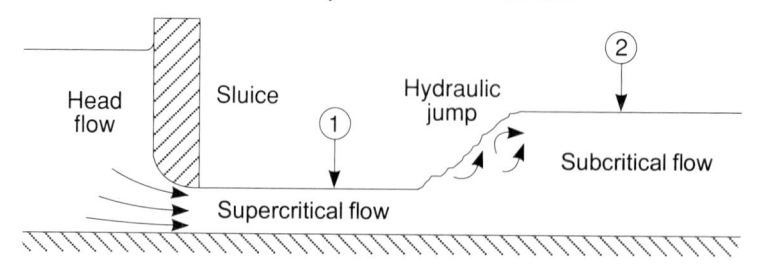

Fig. 8.3 Hydraulic jump

can be neglected, and that the flow is uniform at sections 1 and 2. For depth h and flow area A, we obtain (see for instance (**19**)):

Momentum decrease from 1 to 2 = force due to pressure opposing flow

$$V_2^2 A_2 - V_1^2 A_1 = \tfrac{1}{2}g(h_1 A_1 - h_2 A_2) \tag{8.6}$$

Continuity requires that $V_1 A_1 = V_2 A_2$ and since $A = bh$

$$V_2 = \frac{V_1 h_1}{h_2}$$

if b is constant, and equation (8.6) becomes

$$V_1^2(h_2 - h_1)h_1 = \tfrac{1}{2}gh_2(h_2 - h_1)(h_2 + h_1)$$

and has two solutions:

(i) $h_2 = h_1$ which is the solution for flow without loss

(ii) $V_1^2 = \dfrac{gh_2}{h_1}\dfrac{h_2 + h_1}{2}$ with the solution

$$h_2 = -\tfrac{1}{2}h_1 + \tfrac{1}{2}h_1\sqrt{\left(1 + \frac{8V_1^2}{gh_1}\right)} \tag{8.7}$$

The hydraulic jump is analogous to the shock wave in a gas, requiring supercritical flow and being followed by subcritical flow, and with the depth increasing through the jump, as the pressure increases through the shock. One of the most famous examples of an hydraulic jump is the Severn Bore.

CHAPTER 9

Multiphase Flow

In this chapter the following are covered:
- flow from an oil well
- horizontal two-phase flows
- gas in solution
- cavitation
- other multiphase flows
- flow maps
- gas entrapment
- bubbles, drops and particles

9.1 TYPES OF FLOW

The term multiphase flow is somewhat misleading as it covers both multicomponent and multiphase. Thus dirty gas flows, air-in-water, cavitation, and steam, may all be referred to as multiphase flows within this wide terminology. Because of the dearth of data the fluid engineer attempts to learn from different sources. Below are some important types of flow which generally fall within this definition.

9.2 FLOW FROM AN OIL WELL

We may consider first the flow from an oil well, making the assumption that this is vertical. This is not strictly true, but will give us some basic concepts. This is described in reference (**36**). The crude oil will reach the well-head having flowed up a pipe of about 100 mm bore for distances of several kilometres or about 30 000 pipe diameters. The flow will, therefore, presumably, be fully developed. Initially, when the well is new, the flow will be single phase, essentially oil only. As oil is removed from the well the pressure in the reservoir will decrease and the gas fraction in the well flow

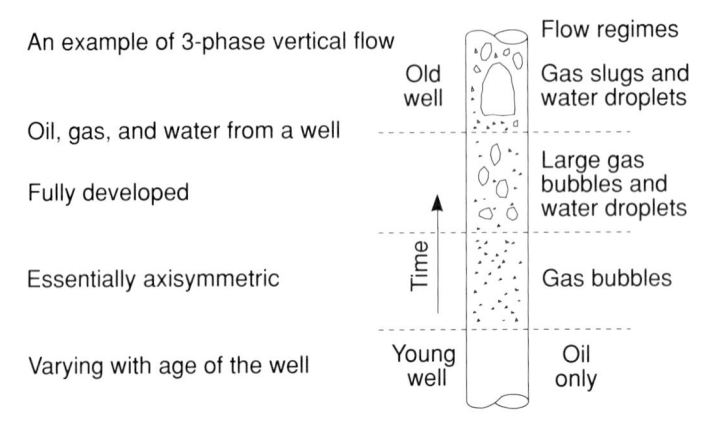

Fig. 9.1 An example of three-phase vertical flow

Labels (left to right): An example of 3-phase vertical flow; Oil, gas, and water from a well; Fully developed; Essentially axisymmetric; Varying with age of the well. Centre: Old well (top) to Young well (bottom), Time axis. Flow regimes (top to bottom): Gas slugs and water droplets; Large gas bubbles and water droplets; Gas bubbles; Oil only.

line will increase and appear as gas bubbles. This is known as bubbly flow (Fig. 9.1). With further ageing and depletion of the reservoir, the gas bubbles will become larger. An equilibrium size distribution will result from breakup due to turbulence, and coalescence due to the breakdown of the liquid film on close approach of two bubbles. In addition the water content is likely to increase. Water will be present as droplets and will form a third phase. Yet further ageing will result in slugs of gas which travel up the centre of the pipe leaving a slower layer of moving liquid on the wall (which, at times, may even reverse in direction during the passage of the slugs). These slugs tend to overtake each other forming larger slugs many metres long. These slugs may be up to 20 m long and may form an equilibrium size distribution giving a balance between coalescence due to bubbles overtaking each other, and breakup due to instabilities when they become too large. As a further complication, the oil flow may contain waxy deposits, and sand. While separation of components is a standard process, the development of subsea systems may require multi-component handling.

9.3 HORIZONTAL TWO-PHASE FLOWS

Horizontal two-phase flows are considered next (Fig. 9.2). The most obvious effect is the loss of axisymmetry. Gravity now causes the less

Multiphase Flow 137

| Single phase | Bubbly | Plug |

| Slug | | Stratified |

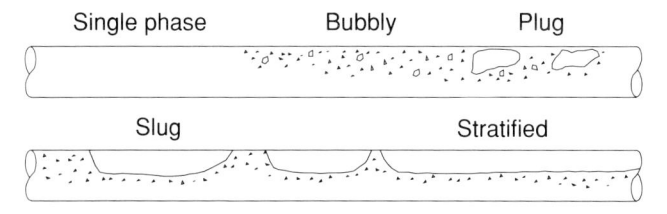

Some regimes are shown above.

Fully developed flow, 100 D or more is needed.

Flow pattern is *not* axisymmetric.
To increase uniformity mixing may be used and a bubbly or droplet distribution will result temporarily.

Fig. 9.2   Horizontal two-phase flow

dense phase to migrate to the top of the pipe. Thus, in a gas/liquid flow the gas will move to the top of the pipe as bubbles. If these are allowed to become large, plugs of gas result, and as these coalesce slugs of gas take up regions against the top of the pipe. Eventually a sufficient number of these will lead to stratified flow.

Alternatively the mixtures may be of two liquids such as water in oil. The droplets of water will sink towards the bottom of the pipe, mirroring the behaviour of air bubbles, and will eventually drop out onto the bottom of the pipe causing a continuous layer of water. Figure 9.3 shows a plot of the distribution of water concentration, C, in oil across a commercial pipeline compared with the mean concentration, $\bar{C}$. The equation for deducing this may be obtained from (**20**). The distribution shown results from mixing in the pipeline upstream.

With a sufficient length of straight pipe (100D or more) a fully developed flow may be achieved. However, although flow pattern maps have been made to predict the nature of the flow in a pipe, the many parameters affecting the flow and the unlikelihood of adequate pipe length to give fully developed flow make it likely that in most applications we shall not be able to predict the resulting flow or how it will affect equipment in the line. This presents a considerable challenge to the development of equipment and instrumentation for handling oil-well

138 An Introductory Guide to Industrial Flow

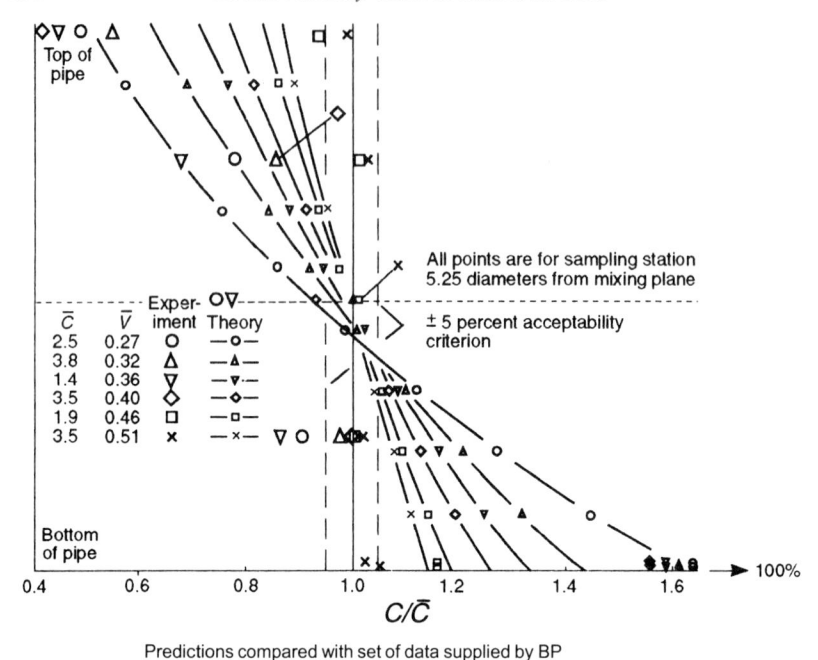

Fig. 9.3 Concentration distribution from Baker (20)

Predictions compared with set of data supplied by BP
(ρ = 828 kg/m^2, v = 5.3 cSt)

All points are for sampling station 5.25 diameters from mixing plane

± 5 percent acceptability criterion

$\bar{C}$	$\bar{V}$	Experiment	Theory
2.5	0.27	○	—○—
3.8	0.32	▲	—▲—
1.4	0.36	▽	—▽—
3.5	0.40	◇	—◇—
1.9	0.46	□	—□—
3.5	0.51	×	—×—

C/$\bar{C}$

flows in subsea installations (**37**). It may be possible to mix the fluid to create a more homogeneous fluid for flow measurement, but this will cause severe turbulence and a changing profile, conditions generally considered unsuitable for such a measurement.

9.4 GAS IN SOLUTION AND AIR ENTRAINMENT

Liquids may contain gases in solution. For water the maximum amount is about two percent by volume at atmospheric pressure. The gas in solution does not increase the volume of the liquid by an amount equal to its volume since the gas molecules 'fit' in the 'gaps' in the liquid molecular structure.

For hydrocarbons the amount of gas which can be held in solution is very large and the GOR (gas-oil-ratio), which is the volume of gas at standard conditions to the volume of liquid, can range up to 100 or more. In either case, but particularly the latter, changes in flow conditions, for instance a pressure drop, can cause the gas to come out of solution causing a two-phase flow.

In low head flows, for instance in a sewage works where some of the flow is in open channels, the flowing stream may entrain air, and a dispersion of air in water may result.

9.5 CAVITATION

Cavitation may occur in certain liquid flows at pressures around ambient. Cavitation is the creation of vapour cavities within the liquid due to localized boiling at low pressure. It can cause damage, for example in pumps, propellers, valves and other flow components since the cavities can collapse very quickly and erode a solid surface. It can also cause errors in flowmeter readings, since it results in a larger volume than for the liquid alone.

9.6 OTHER MULTIPHASE FLOWS

High humidity creates problems of a similar type if it results in a consequent large amount of water vapour changing to liquid droplets in the gas.

Particulate matter can, in addition, cause wear, and may need to be removed with a fine filter.

One truly two-phase fluid is steam. Its importance merits a separate chapter, but some aspects should be noted here. Superheated steam may be treated as a gas and its properties are well tabulated. However, it is increasingly important to measure the flow of wet steam made up, say, of about 95 percent (by mass) vapour, and about 5 percent liquid. The droplets of the liquid are carried by the vapour, but will not follow the vapour stream precisely. As with water droplets in oil the liquid will drop through the vapour to land on the pipe wall and may result in an annular flow regime until sufficient turbulence is created to re-entrain this liquid. The measurement of such a flow causes major problems, since the pressure and temperature remain constant while the dryness fraction changes. It is, therefore, not possible to deduce the dryness fraction or density from the pressure and temperature, and in addition the droplets will cause an error when an attempt is made to measure the flow.

9.7 FLOW MAPS

These complex flows have been extensively studied, and flow pattern maps have been developed to indicate the conditions under which the various flow regimes occur (**38**) (**39**). Figure 9.4 indicates the main features for an imaginary fluid combination. The superficial velocity shown on each axis is that which the gas or liquid would have if it filled the whole pipe. The scales used on the axes may be logarithmic to allow for a large range of values. Thus, referring to the map, at low flows the gas flows above the liquid, causing a stratified flow. As the liquid velocity increases the gas is likely to be broken up into bubbles, but if the gas velocity also increases,

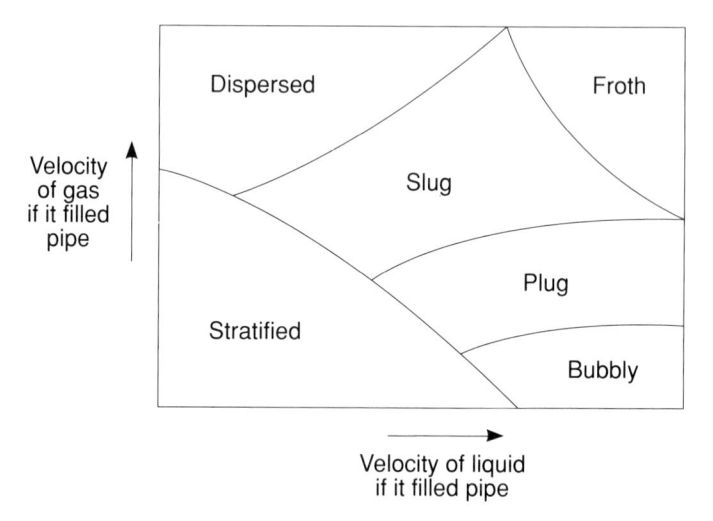

Fig. 9.4 Example of a flow pattern map

then the more violent plug, slug, and froth flows may result. If the gas flow is high and the liquid flow is low, the droplets will be dispersed through the turbulent pipe flow.

Such flow maps tend to be limited in their application to particular fluid combinations and pipe sizes. The handling of these flows in pipes, pumps, valves, and meters may need a better understanding of their nature than such a map can offer. To measure the flow in any of these regimes requires a large number of variables to be interpreted from measurements. Mixing or conditioning of the flow may be possible in some applications, but may not greatly improve the situation. Separation of the components occurs in some applications, but again this is not always possible or convenient.

9.8 GAS ENTRAPMENT

Gas entrapment appears to occur in some flow geometries with important effects. Thomas *et al.* (**40**) observed that "Transient large eddies (vortices) in turbulent free shear flows entrap and transport large quantities of bubbles, and may also force the coalescence of bubbles."

At the conference where this paper was presented the appearance of this phenomenon was discussed in relation to two papers in particular which suggested the presence of the phenomenon in flowmeters. In one of these a turbine meter had been tested (**41**) which appeared to exhibit hysteresis in its response to increasing and decreasing fractions of air in water. This may have been due to the particular meter or the flow circuit, but it raised speculation as to whether vortex structures in the vicinity of the fluid access path to the ball bearing could be entrapping air and holding it in the bearing after the external air content in the flow had dropped.

In another paper (**42**) the entrapment of a second phase had been observed in the shed vortices behind a meter bluff body (Fig. 9.5).

The author has also observed, in some field data from an ultrasonic flowmeter, a behaviour which could result from entrapment of air. A meter in a low head flow, where air was entrained with the water, periodically failed. The possibility that in such a flow the transducer cavities could cause small local vortices which would entrap the air and block the ultrasonic beam offered a possible explanation.

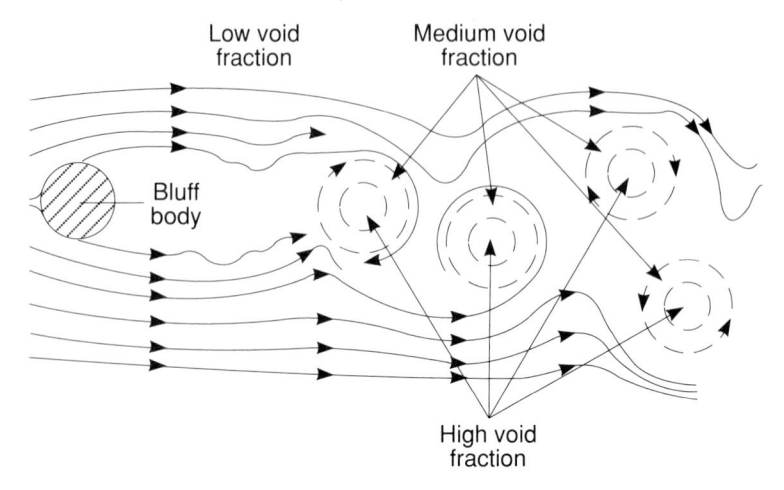

Fig. 9.5 Gas trapped in vortices

There is considerable scope for investigating the appearance of such entrapment and the likely effect in other applications. In flow measurement, entrapment may occur: downstream of an orifice plate; around the throat of a Dall tube; behind an averaging pitot tube. In all these cases it is likely to affect the reading of the flowmeter. Entrapment may also occur in pumps, where it may reduce performance, and in pipework where abrupt area changes exist.

9.9 BUBBLES, DROPS AND PARTICLES

In many cases bubbles, drops and particles in a liquid or gas continuum move relative to the continuum due to gravity or accelerating and decelerating flows. It is, therefore, useful to have a simple method of obtaining the forces on such particles from which their trajectory within the flow can be deduced. For very low Reynolds numbers a drag coefficient for spherical particles due to Stokes is given by

$$C_D = 24/Re \tag{9.1}$$

where the Reynolds number is based on the diameter of the particles

$$Re = \frac{\rho V d}{\mu}$$

Equation (9.1) can be compared with experimental data given in Table 9.1 (**43**)

Table 9.1 Comparison of drag coefficients for spherical particles from Morsi and Alexander (43) with Stokes Law

Reynolds number Re	C_D Experimental values	C_D Stokes Law	Error percentage
0.1	240	240	–
0.2	120	120	–
0.3	80	80	–
0.5	49	48	2
0.7	36.5	34.3	6
1.0	26.5	24.0	9
2.0	14.4	12.0	17
3.0	10.4	8.0	23
5.0	6.9	4.8	30

Clift *et al.* (**44**) showed that for non-spherical particles of most shapes the drag is greater than for a sphere and the only case where it decreases, that of spheroids, the decrease is less than 5 percent. The case of droplets in a liquid is rather more complicated as drag at the surface of the droplet can cause internal circulation.

The terminal velocity for a spherical particle is given by (Fig. 9.6)

$$W = \frac{g d^2 \Delta \rho}{18 \mu} \quad (9.2)$$

where $\Delta \rho = \rho_p - \rho$ is the density difference between the particles and the bulk fluid. If circulation takes place, an additional factor is added of

$$f_{\text{circ}} = \frac{3(1+\kappa)}{2+3\kappa} \quad (9.3)$$

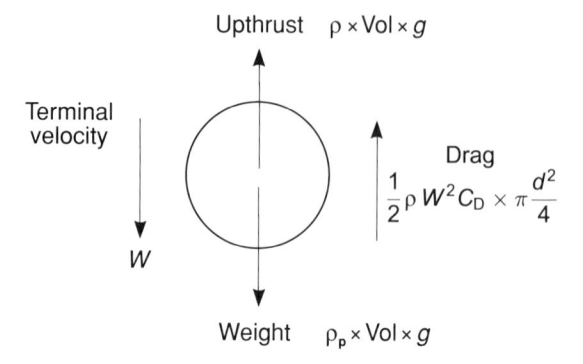

Fig. 9.6 Forces on a droplet falling through another fluid

(Diagram: droplet with Upthrust $\rho \times \text{Vol} \times g$ upward, Weight $\rho_p \times \text{Vol} \times g$ downward, Drag $\frac{1}{2}\rho W^2 C_D \times \pi \frac{d^2}{4}$ upward, Terminal velocity W downward)

so that

$$W = \frac{gd^2 \Delta \rho f_{\text{circ}}}{18\mu} \quad (9.4)$$

where $\kappa = \mu_p/\mu$ and tends to infinity for solids, thus resulting in a factor of unity. For water droplets in oil where $\kappa \ll 1$, the terminal velocity may increase by 50 percent over that for a particle. It has been found that contaminants may alter the behaviour of the fluid interface and retard the surface motion and hence the recirculation, so that W approaches the value for solid particles (**44**).

Example

Estimate the distance along the pipe for a water droplet to fall from the top to the bottom of the pipe, for an oil flow in a horizontal pipe of 1 m diameter flowing at 2 m/s with oil of viscosity $\mu = 0.009$ Pas ($= 9$ cP) and density, $\rho = 800$ kg/m^3

for a droplet of water of 1 mm diameter (μ = 0.001 Pas = 1 cP and ρ_p = 1000 kg/m³)

$$W = \frac{9.81 \times (0.001)^2 \times 200}{18 \times 0.009} = 0.012 \text{ m/s}$$

$$f_{circ} = \frac{3(1+\kappa)}{2+3\kappa} = 1.43$$

Transit time for 1 m:
- without droplet circulation = 1/W = 83 secs;
- with droplet circulation = 1/(f_{circ}W) = 58 secs.

For a flow of 2 m/s this will result in the droplet moving down the pipe between 116 m and 166 m.

The effect of turbulence has been ignored and the problem of how to determine the droplet size is not addressed here but the reader is referred to reference (**20**).

Similar effects to those of gas-entrapment occur when fine particles are carried by flows, and lead to deposition in recirculation regions. This may be observed as dust deposits where such regions exist, as, for example, on the walls of underground tunnels.

Examples of the trajectories of particles in water are given in the final chapter, and the technology of cleaning by suction raises many interesting problems relating to the entrainment of particles.

CHAPTER 10

Steam

In this chapter we shall cover:

- some thermodynamic definitions;
- how to use tables and charts;
- some equations of state;
- some simple flow equations;

and we shall look at some examples using the tables and chart.

10.1 INTRODUCTION

What happens when water is heated? It is useful to imagine the effect of so doing as in Fig. 10.1. Imagine a vertical cylinder containing one kilogram of water. On top of the water there is a piston sealing on the walls of the cylinder and creating a water pressure of 10 bar.

A   The water is at ambient temperature. We find that by heating the water we raise the temperature but cause hardly any volume change and no vapour.
B   The water has been heated to the limit at which only water is present. Any further heating will cause no further temperature change, but will cause vapour to appear.
C   Further heating has taken place with no change in temperature and now, under the piston, is a two-phase mixture of liquid and vapour.
D   The limit has been reached where all the water has evaporated and there is only vapour present. Any further heating will now cause an increase in temperature again.
E   The vapour is now superheated and the volume of the vapour under the piston is related directly to the temperature (the pressure being kept constant).

Pressure

P_c = 221.2 bar
T_c = 374.15°C

Critical point

Liquid region — Two-phase region

Gaseous region

A B   C   D   E

Saturated liquid line

Saturated vapour line

Specific volume ⟶

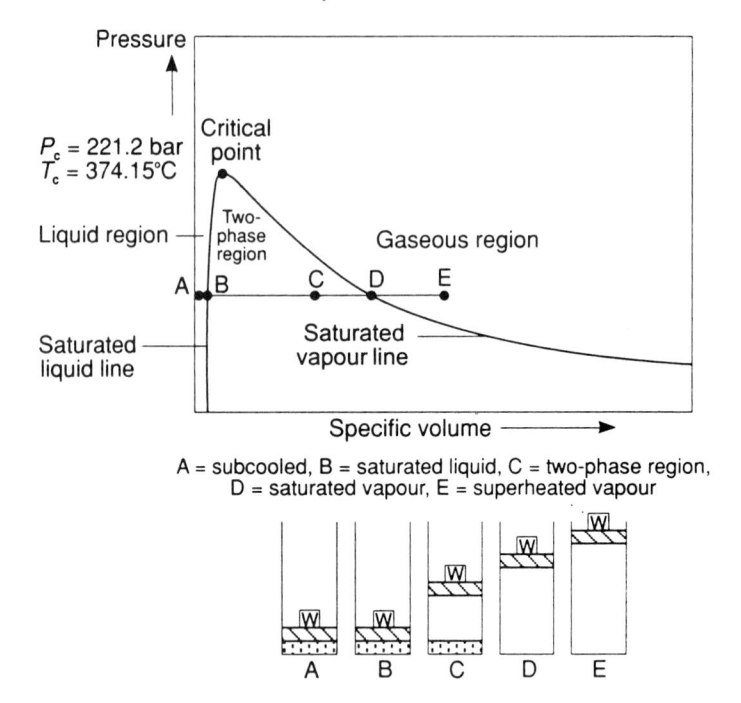

A = subcooled, B = saturated liquid, C = two-phase region,
D = saturated vapour, E = superheated vapour

Fig. 10.1 Diagram illustrating the behaviour of steam (after Budenholzer) (45)

These changes are plotted on the pressure/specific volume diagram and it can be seen that B and D lie on the saturated liquid line and the saturated vapour line, respectively. The fluid in regions A and E behaved differently from that in region C. In A and E a change in volume was directly related to a change in temperature at constant pressure. In region C another factor entered – the amounts of liquid and vapour. Indeed at C a knowledge of temperature and pressure was not sufficient to define the point on the line BD. In region E the vapour has a gas-like behaviour and we shall calculate how good this approximation is later.

Before looking more closely at these changes some definitions will be useful. The table and chart will then be introduced. Finally there is a brief review of some important ideas and equations which are needed to calculate simple flows and the chart will be used to do some examples.

10.2 DEFINITIONS

The following definitions are quoted from Keenan (**22**). Some of them were introduced in Chapter 3.

A system
A system is any collection of matter enclosed within prescribed boundaries.

A pure substance
A system which is homogeneous in composition and homogeneous and invariable in chemical aggregation is called a pure substance. It is known from experience that a pure substance in the absence of motion, gravity, capillarity, electricity and magnetism has only two independent properties.

Thus water as vapour, liquid, and ice is a pure substance, but a mixture of hydrogen, oxygen, and water is not.

A property
A property of a system is any observable characteristic of the system.

Within the two-phase region for water, pressure and temperature are clearly not independent properties. So a second property is needed, such as dryness or density to define the substance.

A state
A state of a system is its condition or position and is identified through the properties of the system.

Examples of extensive properties which are additive are

- v specific volume
- u internal energy
- h enthalpy

while other properties which are not additive are

- T temperature
- p pressure
- ρ density

If r is taken as any extensive property for unit mass of a two-phase mixture

$$r = xr_g + (1-x)r_f \tag{10.1}$$

where x is the mass fraction of vapour and the subscript f refers to saturated liquid and g to saturated vapour. Thus if $x = 0$ there is no vapour in the mixture and $r = r_f$, while if $x = 1$ there is no liquid in the mixture and $r = r_g$.

Thus v would behave as r and the total volume per unit mass will be the sum of the volume fractions of liquid and vapour. ρ will be obtained as the reciprocal of v. On the other hand, temperature will be the same for both the liquid and vapour phases.

10.3 TABLES AND CHARTS

In Fig. 10.2 the saturation line is plotted on a p–v diagram. It is clear that the value of the temperature will not vary for constant pressure in the two-phase region, but outside that region the curves of constant pressure (isobars) and constant temperature (isotherms) do not coincide. Figure 10.2 enables us to obtain the temperature necessary to obtain steam of a certain pressure and specific volume in the superheated region. If the pressure and specific volume indicate that the state of the system is in the two-phase region, then the values of v_f and v_g are found from the points at

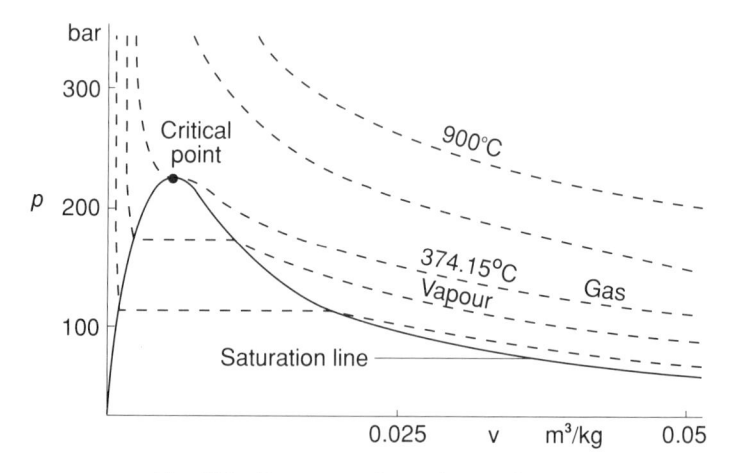

Fig. 10.2 Pressure-volume diagram for steam

which the horizontal pressure line intersects the saturation line, and the dryness is then obtained from equation (10.1).

An alternative diagram is one of pressure against temperature. It is notable that the two-phase region, where there is one value of pressure for one value of temperature, will collapse on to a line. This is shown in Fig. 10.3.

It is of interest to note that increased pressure in the ice region can cause melting at constant temperature (useful for skating as it provides a liquid lubrication layer between the skate and the ice). The letters designate regions which are the same as in Fig. 10.1. The table in Appendix B covers part of A, B, C, D and E for the density of water and steam.

A three-dimensional diagram of the surfaces of the p–v–T graphs is shown in Fig. 10.4 and has some interesting features. Figure 10.3 can be found by looking along the triple point line in the negative v direction. One can also see that as the liquid freezes the specific volume increases. This is the case for water. Figure 10.4 is based on a diagram from Zemansky (**46**) who defined a gas as a substance with temperature above the critical isotherm.

Tables are commonly available for the saturation region which is shown as a single curve in Fig. 10.3. These are either with temperature or pressure

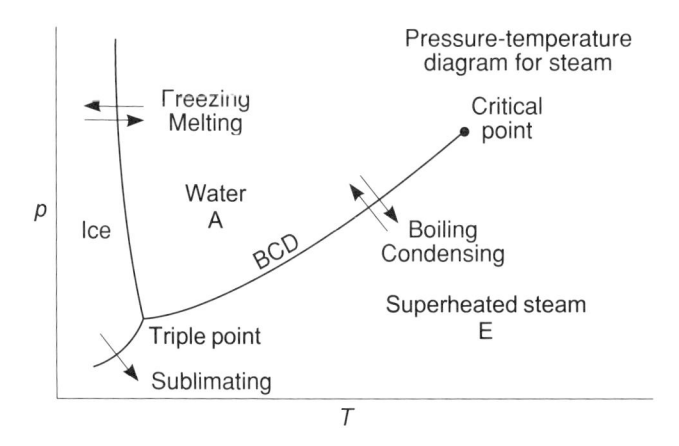

Fig. 10.3 Pressure-temperature diagram for steam

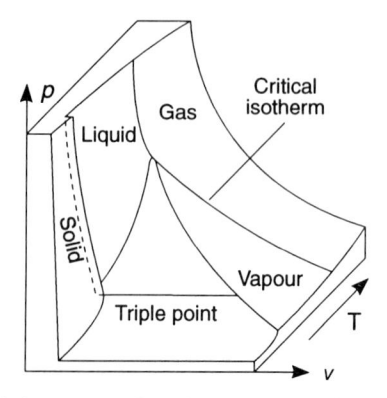

Fig. 10.4 p-v-T surface for water (after Zemansky) (46)

as independent variable with maximum values of

$p = 221.2$ bar
$T = 374.15°C$

for the critical point. These tables will give values of v_f, v_g, u_f, u_g, h_f, h_{fg}, h_g, s_f, s_g or a similar list.

A table covering the region for water is sometimes combined with the region for superheated steam and one table then allows ρ, for instance, to be obtained for values of p and T. Such a table is reproduced at the end of this book as Appendix B. Similar tables may be obtained for u, h and s (cf steam tables). The line dividing the region for water from the region for superheated steam is then shown in the table and saturation values on the line may also be included. The triple point (the point where ice, water and steam can coexist) is at $T = 273.16$ K and 0.6112 kPa (**47**).

Keenan (**22**) gives the following notes on the critical point.

"The critical state may be observed by filling a quartz tube with the proper proportions of a liquid and its vapour and then heating the tube and its contents. ... The meniscus separating the two phases will remain near the middle of the tube until, as the critical pressure is approached, it becomes less distinct and finally disappears. At pressures higher than the critical pressure the liquid can be heated

Steam

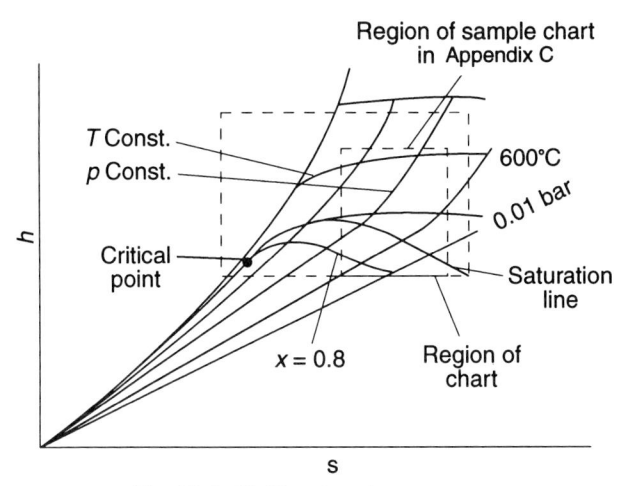

Fig. 10.5 Mollier chart for water

from a low temperature to a high one without any discontinuity in the process. Ebullition does not occur as at lower pressures and no other event marks a change...."

Oil wells at greater depths and, therefore, greater pressures are, in some cases, supercritical, and may raise new problems of handling the fluid.

Other plots are useful for various purposes, but for the present purposes we will turn to the chart of enthalpy–entropy, a sample of which is supplied with this book (**48**). The importance of this chart will become apparent when the main equations are considered. Now we consider briefly the chart's structure. In Fig. 10.5 its relation to the whole of the h-s diagram is shown. The saturation line is shown, below which liquid and vapour coexist and the 80 percent dryness curve can be seen. Constant pressure curves are shown stemming from the origin of the axes and within the two-phase region these are collinear with the constant temperature curves. Outside the two-phase region the constant temperature curves tend to the horizontal and the 600°C line is shown.

10.4 EQUATIONS OF STATE

> **Values of the gas constant, R, for steam**
>
> **From steam tables**
>
> $T = 400°C = 673$ K $\quad\quad \rho = 0.322$ kg/m^3
> $p = 1$ bar $= 10^5$ N/m^2 $\quad\quad \therefore R = 461.45$ J/kgK
> $T = 800°C = 1073$ K $\quad\quad \rho = 2.02$ kg/m^3
> $p = 10$ bar $= 10^6$ N/m^2 $\quad\quad \therefore R = 461.37$ J/kgK
>
> **From the universal gas constant**
>
> $\mathbf{R} = 8314.3$ J/kmolK
> $\mathbf{M} = 18$ $\quad\quad\quad\quad\quad\quad R = 461.91$ J/kgK
>
> This approximate behaviour is calculated below

For pressures well below the critical pressure (p_c) and temperatures well above the critical temperature (T_c) the relation for a gas is approximated by

$$pv = RT = \frac{\mathbf{R}T}{\mathbf{M}} \tag{10.2}$$

where R is the gas constant for a particular gas. $\mathbf{R}$ is the universal gas constant and $\mathbf{M}$ is the molecular weight for a particular gas. From the steam tables for

$T = 400°C = 673$ K
$p = 1$ bar $= 10^5$ N/m^2

we obtain

$$\rho = \frac{1}{v} = 0.322 \text{ kg/m}^2$$

and for

$T = 800°C = 1073$ K
$p = 10$ bar $= 10^6$ N/m^2

we obtain

$$\rho = \frac{1}{v} = 2.02 \text{ kg/m}^2$$

Using the same units in each case

$$R = \frac{10^6}{673 \times 0.322} = 461.45 \text{ J/kgK}$$

$$R = \frac{10^6}{1073 \times 2.02} = 461.37 \text{ J/kgK}$$

Using **R** = 8314.3 J/kmolK and **M** = 18 we obtain

$$R = 461.91 \text{ J/kgK}$$

Thus the assumption that steam is an ideal gas at these conditions is within about 0.1 percent.

Van der Waal's equation attempts to allow for finite size molecules which reduce the volume, and inter-molecular attraction which alters p. The equation is

$$p = \frac{RT}{v-b} - \frac{a}{v^2} \tag{10.3}$$

and this is plotted in Fig. 10.6 with the curve of maxima and minima. The two-phase region will extend outside this curve except at the critical point, such that it intersects each constant temperature line at the same pressure on each side. Thus Van der Waal's equation suggests that the constant temperature lines will continue to fall after crossing the saturation line from liquid, and will continue to rise after crossing the saturation line from vapour.

From the figure it is apparent that the critical point is a point of inflection and this means that the curve is momentarily horizontal at this point and that the change in the slope at this point is also momentarily

156   An Introductory Guide to Industrial Flow

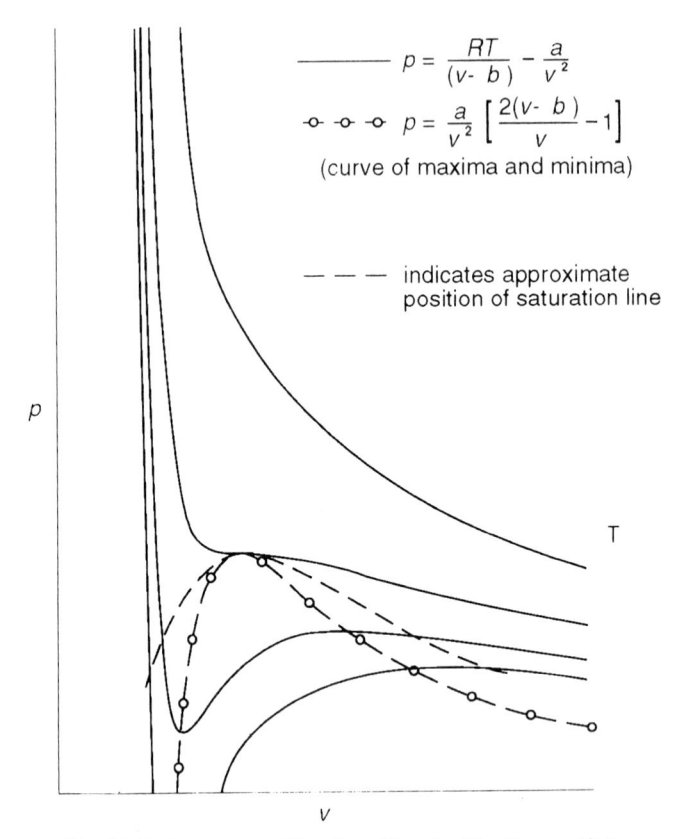

```
——————  p = RT/(v-b) - a/v²
-o- -o- -o-  p = a/v² [ 2(v-b)/v - 1 ]
              (curve of maxima and minima)

— — —  indicates approximate
        position of saturation line
```

Fig. 10.6 Curves resulting from Van der Waal's equation

zero. Mathematically this is written as

$$\left(\frac{\partial p}{\partial v}\right)_T = 0 \tag{10.4}$$

and

$$\left(\frac{\partial^2 p}{\partial v^2}\right)_T = 0 \tag{10.5}$$

With these equations it is possible to show that

$$v_c = 3b$$

$$p_c = \frac{a}{27b^2}$$

$$T_c = \frac{8a}{27Rb}$$

and hence that

$$\frac{p_c v_c}{RT_c} = \frac{3}{8} = 0.38$$

Zemansky (46) gives actual values of: 0.22 for water; 0.29 for carbon dioxide, oxygen, argon, nitrogen; 0.32 for helium; 0.33 for hydrogen.

One of the interesting points about this equation, although it is not very accurate, is that it does indicate some of the effects which may be present in a real fluid and how these effects will modify the equation. It is known that under some circumstances of high purity it is possible to obtain supersaturated or supercooled vapour and supersaturated or superheated liquid. This appears to be possible because the breakdown from the pure vapour or liquid state to the two-phase state requires particles to cause nucleation. One example of this is the Wilson cloud chamber used to observe the passage of particles of nuclear radiation. To quote Keenan:

> "Wilson ... showed that dust-free air saturated with water vapor can be expanded isentropically to a volume 25 per cent greater before condensation occurs, whereas ... for stable mixtures condensation should begin with expansion".

The passage of a particle is then sufficient with the vapour in this unstable state to cause condensation. Similar effects can be observed for the flow of steam through a convergent–divergent nozzle. It is also well known that for liquid free from dissolved gases in a tube of sufficient strength, a negative pressure can be achieved in the liquid so that it is in a state of tension.

A more complicated equation is the Beattie–Bridgman equation

$$p = \frac{RT(1-\epsilon)}{v^2}(v+B) - \frac{A}{v^2}$$

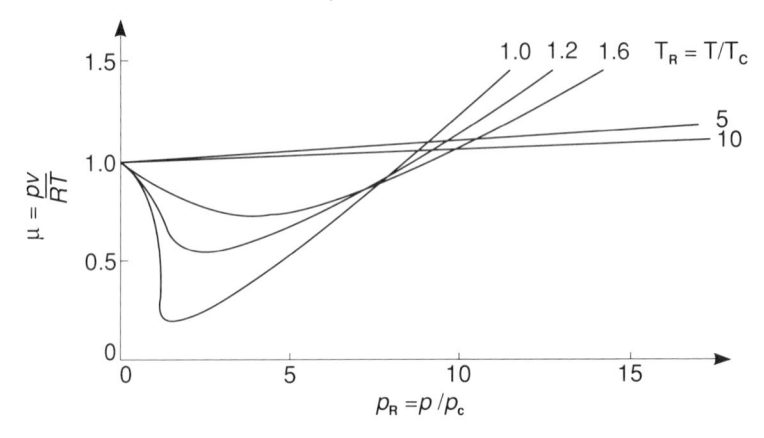

Fig. 10.7 Chart of μ against p_R for various values of T_R

where $A = A_o(1 - a/v)$
$B = B_o(1 - b/v)$
$\epsilon = c/vT^3$

The equations used for properties of steam may also have the form

$$\frac{pv}{RT} = 1 + \alpha_1 p + \Sigma \alpha_i p^i$$

where each α is a function of T only. This is the type of equation used to compute steam tables.

If we write $p/p_c = p_R$, $T/T_c = T_R$, and $pv/RT = \mu$, a graph can be plotted, as shown in Fig. 10.7 which gives an approximation for several gases. For an ideal gas $\mu = 1$, and so this diagram indicates the behaviour of a real gas.

10.5 SOLUTIONS TO SIMPLE FLOW PROBLEMS

The essential equations for flow have been described in Chapters 3 and 6 and are equally relevant here:

- Continuity equation
- Steady flow energy equation

- Reversible flow relationship
- Bernoulli's equation

Example 1 Convergent duct – superheated steam

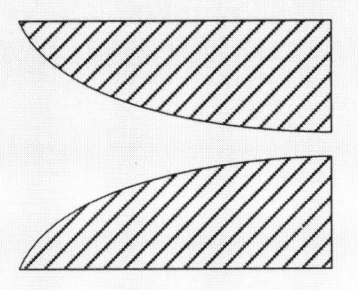

Fig. 10.8 Convergent duct

The flow is considered through the convergent duct in Fig. 10.8. The inlet conditions are 400°C and 10 bar with an inlet diameter of 100 mm. The conditions at the outlet are 9 bar and a diameter of 25 mm. From the chart the inlet enthalpy is obtained as 3263 kJ/kg and from the table in Appendix B the inlet density can be obtained as 3.26 kg/m^3. If it is assumed that the process is isentropic, the enthalpy at outlet can be obtained as 3230 kJ/kg from the chart and the temperature as 384°C. From interpolation in the table we can obtain the density:

Temp (°C)	Pressure (bar)		
	5	9	10
	Density (kg/m^3)		
375	1.685		3.40
384	1.662	3.01	3.35
400	1.620		3.26

The value of velocity at the throat can be obtained as 257 m/s, assuming that the inlet velocity was negligible from the change in the enthalpy 33 kJ/kg. It should also be noted that Bernoulli's equation results from the steady flow energy equation for reversible adiabatic flows. It is interesting to note the value obtained from using Bernoulli's equation (3.28) and incorporating the mean value of density. We obtain 253 m/s. Using the data in Appendix A an approximate value can be obtained for the speed of sound at the duct outlet as $c = \sqrt{(\gamma p/\rho)} = \sqrt{(1.333 \times 9 \times 10^5/3.01)} = 631$ m/s, giving a Mach number of 0.4.

Example 2   Convergent–divergent nozzle: superheated steam

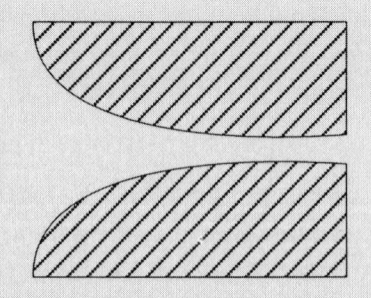

Fig. 10.9   Convergent–divergent nozzle

To consider the behaviour of flow through a convergent–divergent nozzle, Fig. 10.9 is referred to. The initial temperature, $T_1 = 700°C$, and pressure, $p_1 = 50$ bar. The final pressure, $p_2 = 20$ bar. Obtain the outlet velocity, assuming that the inlet velocity is negligible:

(a) if the flow is reversible;
(b) if the efficiency of the process is 80 percent.

The initial state is identified on Fig. 10.10. When the process is reversible the line which follows the initial state is vertical, and the change in enthalpy can be obtained from the intersection of this line with the 20 bar constant pressure line. If equation (3.19) is used for a duct with single entry and exit, $Q = W = 0$, $V_1 = 0$ and $z_1 = z_2$

$$h_1 = h_2 + \frac{V_2^2}{2}$$

The inlet values of enthalpy and entropy from the chart at the end of the book are:

$h_1 = 3900$ kJ/kg
$s_1 = 7.52$ kJ/kgK

(a) For reversible flow the outlet conditions can be obtained knowing p_2 and s_1 as

$s_2 = s_1 = 7.52$ kJ/kgK
$h_2 = 3535$ kJ/kg

$$\frac{V_2^2}{2} = (3900 - 3535) \times 1000 = 365\,000$$

The velocity is, therefore, obtained as

$V_2 = 854$ m/s

Using Appendix A ($\gamma = 1.333$) this value can be compared with one obtained assuming steam to be a perfect gas. With $p/p_0 = 0.4$, $M = 1.244$ and $V_2/\sqrt{(C_p T_0)} = 0.640$ giving $V_2 = 0.640\sqrt{(1921 \times 973)} = 875$ m/s. The throat velocity can be obtained, since for $M = 1$, $V = 0.534\sqrt{(1921 \times 973)} = 730$ m/s.

(b) In an actual flow there will be losses and these will define the efficiency of the flow process:

$$\eta = \frac{h_1 - h_{2'}}{h_1 - h_2}$$

Therefore $h_{2'}$ can be obtained from

$$h_{2'} = h_1 - \eta(h_1 - h_2)$$
$$= 3900 - 0.8 \times 365 = 3608$$

Losses in duct flow for steam

A = Reversible expansion
B = Irreversible expansion

$$\eta = \frac{h_1 - h_{2'}}{h_1 - h_2}$$

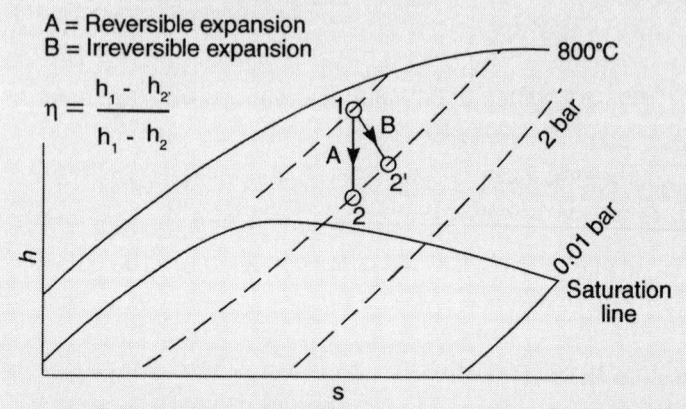

Fig. 10.10 Enthalpy–entropy diagram for steam flow through a convergent–divergent nozzle

Therefore $h_{2'}$ can be plotted on Fig. 10.10, since the enthalpy and the pressure are known. The entropy can also be obtained as

$$s_2 = 7.61 \text{ kJ/kgK}$$

and the new value of velocity at outlet can be obtained as

$$\frac{V_2^2}{2} = (3900 - 3608) \times 1000 = 292\,000$$
$$V_2 = 764 \text{ m/s}$$

We can see that the irreversibility of the process indicated by the efficiency of 80 percent has reduced the outlet velocity by about 10 percent for the same pressure ratio.

Example 3 Saturated steam

In this case the inlet conditions are taken as 200°C and 5 bar and a convergent–divergent duct is assumed with outlet diameter 10 mm. The outlet pressure is now 1 bar. From the chart the enthalpy is obtained as 2858 kJ/kg at entry and from the table the density is 2.353 kg/m^3. Again, we assume an isentropic process and obtain the outlet enthalpy as 2563 kJ/kg and the dryness x as 0.952. Using the equation for extensive properties we obtain

$$v_2 = \frac{0.952}{0.590} + \frac{0.048}{958} = 1.614 \text{ m}^3/\text{kg}$$

So

$$\rho_2 = 1/v_2 = 0.620 \text{ kg/m}^3$$

The enthalpy change is 295 kJ/kg, leading to a velocity at outlet of 768 m/s. Since the cross-sectional area at exit is 0.785×10^{-4} m^2, this gives a mass flow rate of 0.0374 kg/s. To obtain the outlet value of Mach number we can calculate an approximate value for the speed of sound.

Speed of sound

The speed of sound is given by

$$c^2 = (\partial p/\partial \rho)_s$$

For saturated steam the value of the derivative can be obtained approximately by obtaining the change in dryness for a small pressure change around the exit of 1.2 bar to 1 bar (0.2×10^5). For constant entropy, $s = 7.06$ kJ/kgK, the dryness changes from 0.960 to 0.952. Using the values for density from the table for 1 bar of $\rho_f = 958$ and $\rho_g = 0.590$, we obtain using equation (10.1)

$$v = xv_g + (1-x)v_f$$

$$= \frac{0.952}{0.590} + \frac{0.048}{958} = 1.614 = \frac{1}{0.6197}$$

Interpolating for the values for density from the table for 1.2 bar, we obtain $\rho_f = 956$ and $\rho_g = 0.694$ and using equation (10.1) we obtain

$$v' = \frac{0.960}{0.694} + \frac{0.040}{956} = 1.383 = \frac{1}{0.7229}$$

from which we obtain a change of density of 0.1032 kg/m³ and hence

$$c^2 = \left(\frac{\Delta p}{\Delta \rho}\right)_s = \frac{0.2 \times 10^5}{0.1032} = 193\,800$$

Thus sound speed is approximately 440 m/s. The exit Mach number in this example would, therefore, have been 1.75 and the duct would have had to be convergent–divergent.

CHAPTER 11

Computer Fluid Dynamics – Present achievements and future developments

11.1 INTRODUCTION

> Examples of computational fluid dynamics applied to flows:
> - in pipes and fittings;
> - in machinery;
> - in industrial applications;
> - in flowmeters.
>
> Future developments are considered.

At the start of this book some examples of the experimental investigation of flows in a variety of applications were considered. It is, therefore, appropriate to conclude by looking at the computational advances in flow prediction. The development of CFD (computational fluid dynamics) over the past 30 years, the period during which the computer has developed as a powerful and readily available tool, has been enormous and has transformed our ability to predict flows in the most complex of geometries. Flow computation packages are now so accessible that many of the calculations with idealized geometries and perfect fluids, which have been the subject of this book, may seem as out-of-date as the ability to do mental arithmetic in the age of the pocket calculator. Perhaps, in reality, both disciplines are just as important to the ordinary engineer.

Many of the complex flow patterns to be expected within very detailed flow passages are now obtainable from a computer, and the power of these computational packages is likely to increase.

This chapter looks at some examples of the achievements of computing drawn in part from the author's work with his colleagues.

11.2 FLOW IN PIPES AND PIPE COMPONENTS

In Chapter 3 profiles for flow in a pipe were given and in the case of turbulent flow the curve fit used was a power law without physical basis. In a paper in 1976 by Goulas and Baker (**49**) the through flow method of analysis of viscous and turbulent flows was used to obtain the profile in turbulent pipe flow and this is compared with the power law in Fig. 11.1. In the subsequent twenty years prediction methods have become very much more powerful, and a simple problem such as this one should be amenable to any computational flow package.

Damia Torres (**9**) used a time marching, finite area calculation to obtain the flow through a two-dimensional convergent–divergent duct. In Fig. 11.2(a) computed centre-line Mach number variation is compared with experimental results from laser doppler anemometry measurements. It is interesting to note that the particles used in the LDA measurements cannot follow the sudden changes through the shock wave and lead to a substantial disagreement. This is compounded by the approximations in the model in the diffuser. Figure 11.2(b) which shows the outline of the duct shape is a reminder that the pattern of Mach numbers is not as simple as was assumed in Chapter 6.

In 1993 Lenn *et al.* studied the flow downstream of a 'T' junction (**10**). Such flows are of interest in the production and transfer of crude oil, since

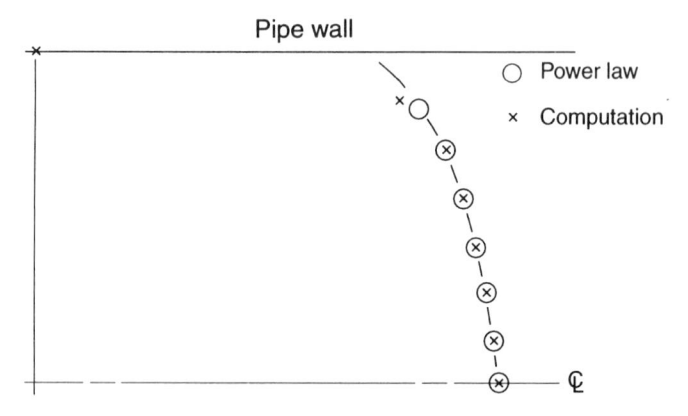

Fig. 11.1 Turbulent flow profile in a straight pipe after 6 diameters at Re = 50 000 (after Goulas and Baker) (46)

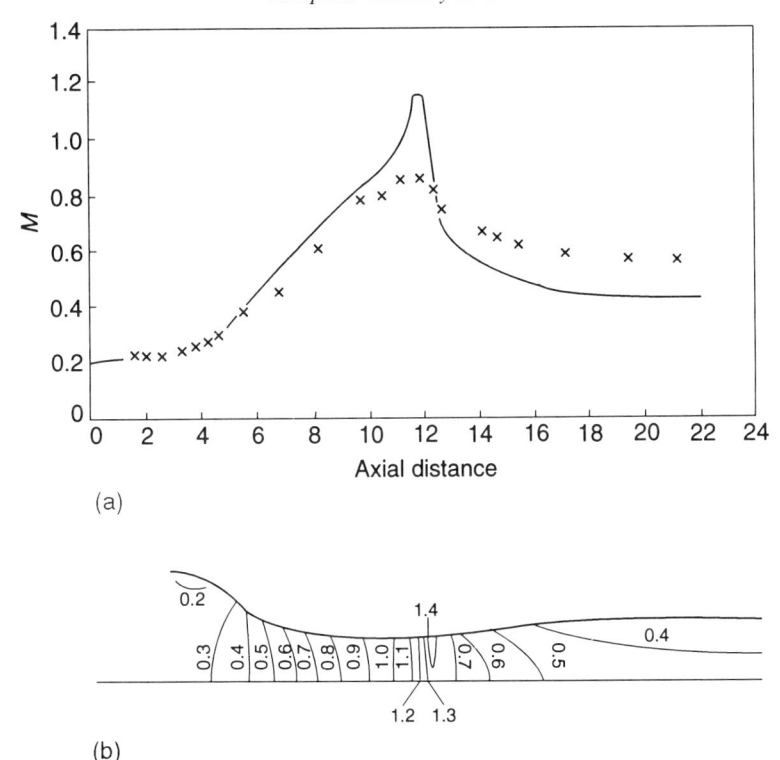

Fig. 11.2 Flow in a convergent–divergent duct (after Damia Torres with permission) (9): (a) Centre-line Mach number — computer X experiment; (b) Computed Mach number contours in the two-dimensional duct

the 'T' junction causes some breakup of droplets due to enhanced turbulence and some mixing of the water in the oil due to the secondary flow patterns. These two effects allow samples of the mixture to be extracted which will be representative of the bulk flow. In Fig. 11.3 the geometry of the 'T' junction is shown and the velocity profiles are reproduced. This work demonstrated the use of mathematics, not to predict the flow profile, but to fit the experimental results of a very complex flow with a family of curves. These are the curves shown in Fig. 11.3 (b)–(d) where the solid and dashed lines are the curve fit of the experimental points which are shown.

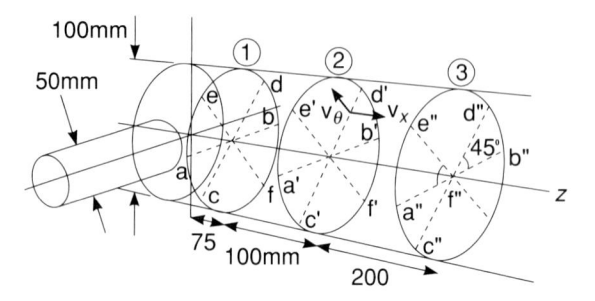

(a) Diameters along which velocity components were measured

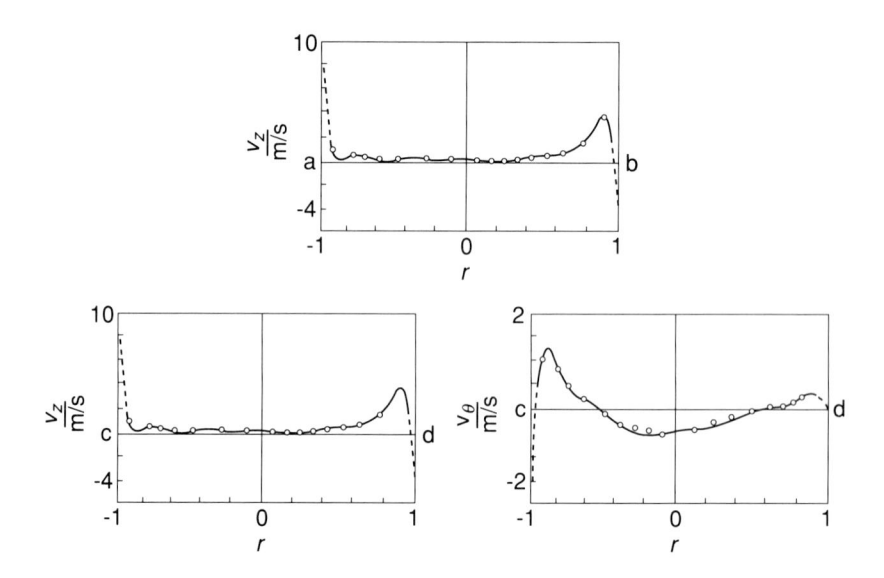

(b) Axial and tangential velocity profiles at station 1 against non-dimensional radius

Fig. 11.3 Flow in a 'T' junction (10)

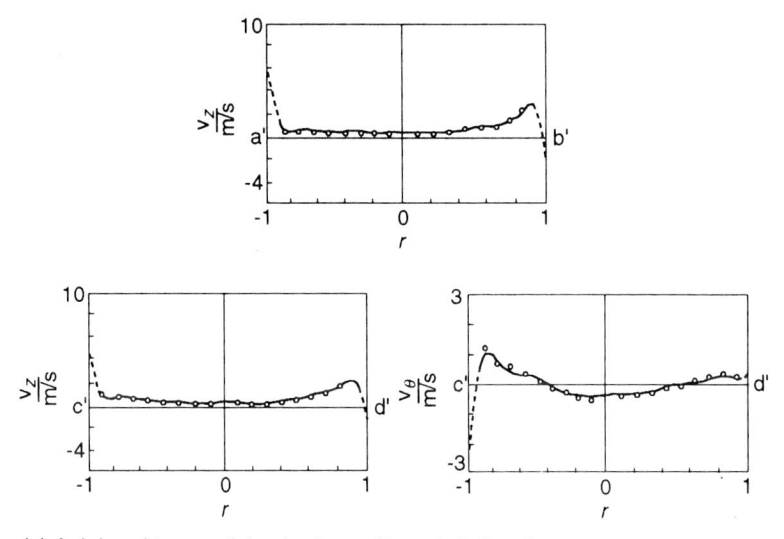

(c) Axial and tangential velocity profiles at station 2

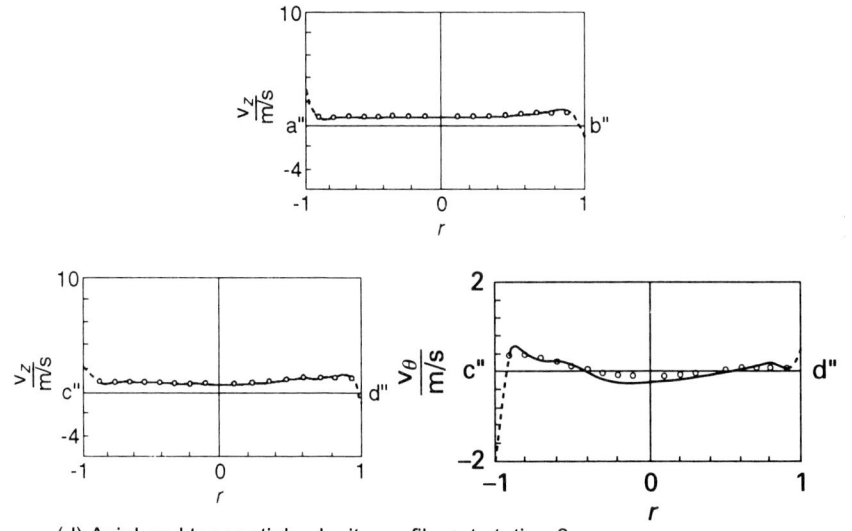

(d) Axial and tangential velocity profiles at station 3

Fig. 11.3 Flow in a 'T' junction (continued) (c) and (d)

170   An Introductory Guide to Industrial Flow

11.3 FLOW THROUGH TURBOMACHINES

Ahmad *et al.* (**4**) used the method described by Goulas and Baker (**49**) to compute the flow patterns in a slurry handling pump, on grid stations, as shown in Fig. 11.4(a) for a cross-section of the radial flow path. Figure 11.4(b) shows the distribution of radial velocities through the pump. With these computations of velocity the trajectories of particles were calculated, examples of which are shown in Figs 11.4(c) and (d). From this erosion-prone areas were deduced as in Fig. 11.4(e) and these can be compared with paint erosion experiments in Fig. 11.4(f). The average experimental erosion pattern is shown in Fig. 11.4(g), and while the agreement between this and the computations is not exact, the general pattern of erosion has similarities.

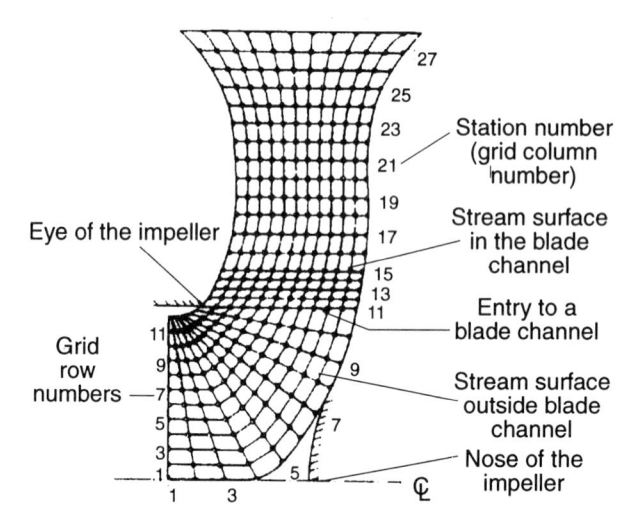

(a) stream surface and the grid mapped out on the surface used for flow calculations

Fig. 11.4 Slurry flow in a pump (4)

Computer Fluid Dynamics 171

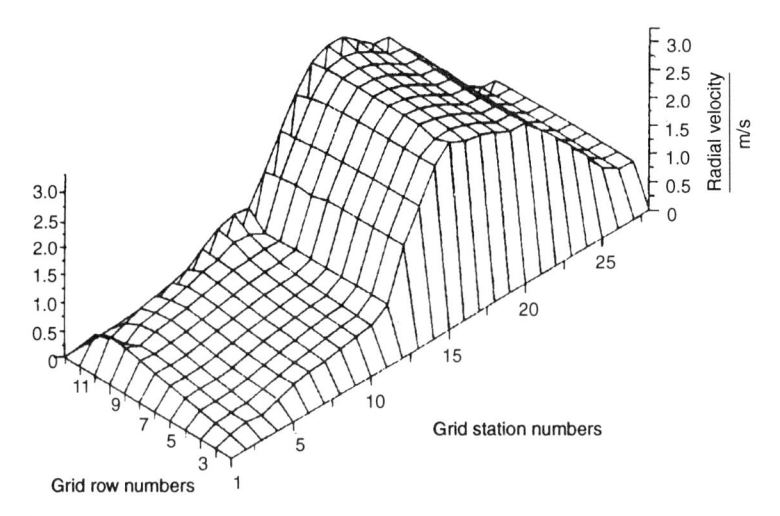

(b) isometric plot of radial velocity along grid stations

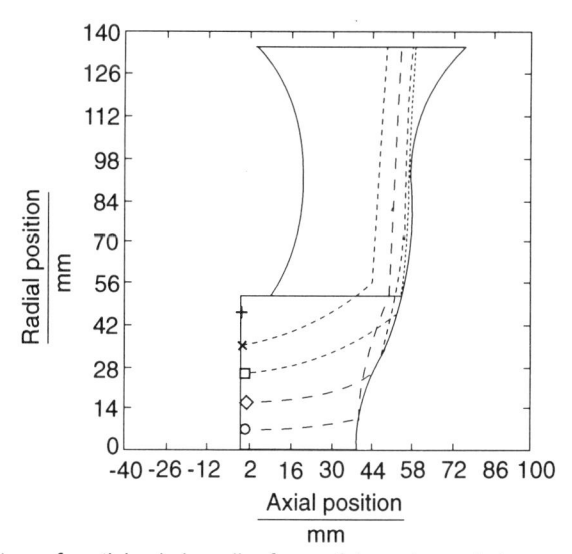

(c) trajectory of particles in impeller for particles released along vertical diameter of pipe

Fig. 11.4 Slurry flow in a pump (4) (continued) (b) and (c)

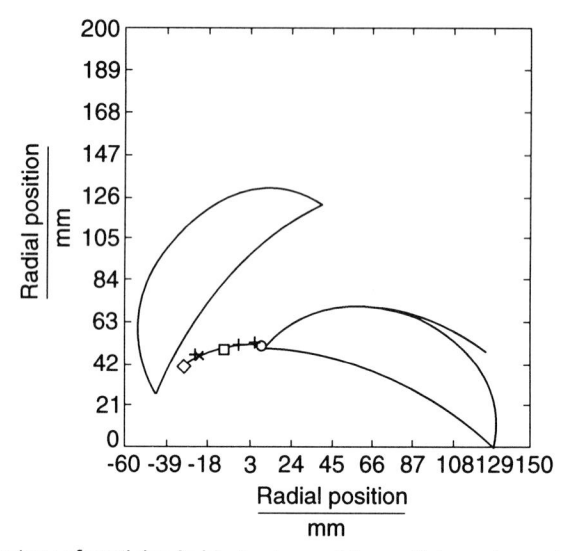

(d) trajectory of particles in blade channel for particles released along vertical diameter of pipe

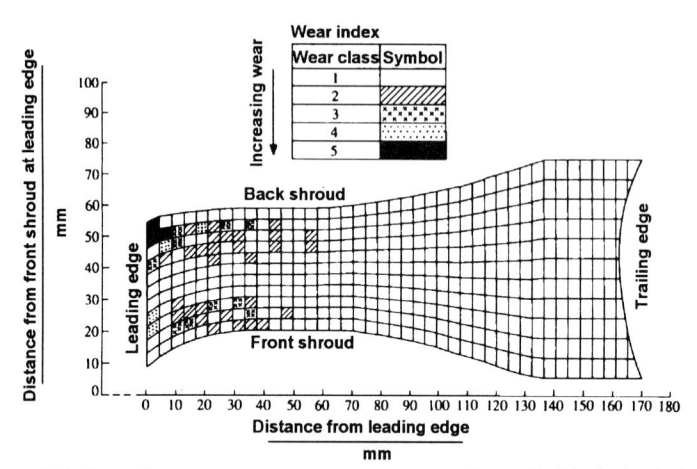

(e) predicted erosion-prone areas on pressure surface of a blade for 3.4 percent concentration and 11.6×10^{-3} m^3/s slurry flow rate

Fig. 11.4   Slurry flow in a pump (4) (continued) (d) and (e)

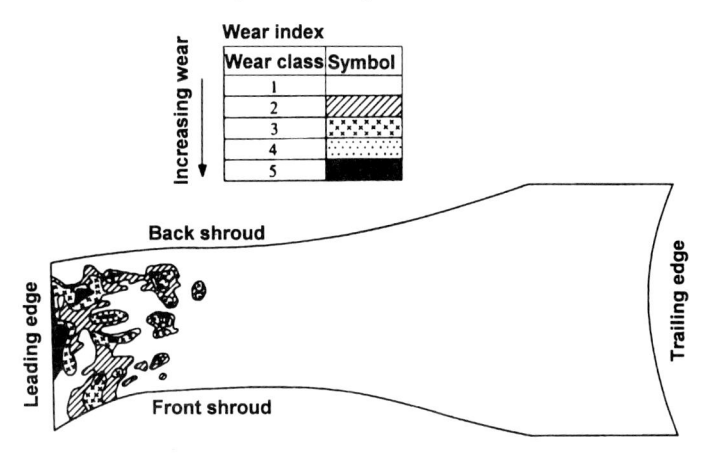

Computer Fluid Dynamics 173

(f) erosion pattern on one blade pressure surface for 3.4 percent concentration

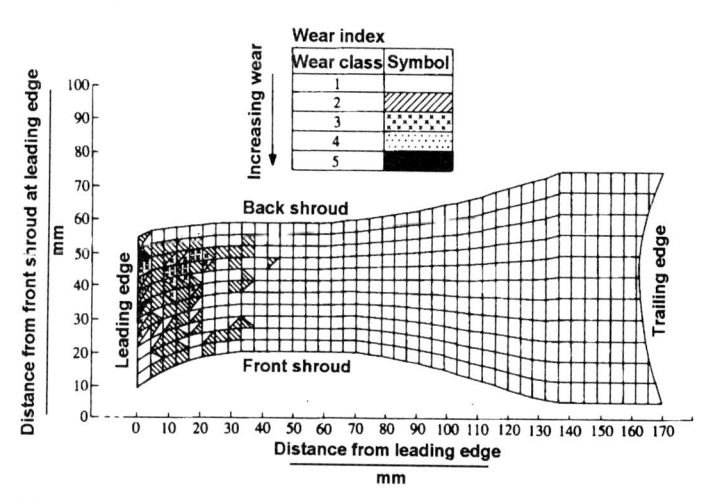

(g) average experimental erosion pattern on a blade pressure surface for 3.4 percent concentration

Fig. 11.4 Slurry flow in a pump (4) (continued) (f) and (g)

(a) geometry of poppet valve

(b) simulated streamlines for z = 0.05

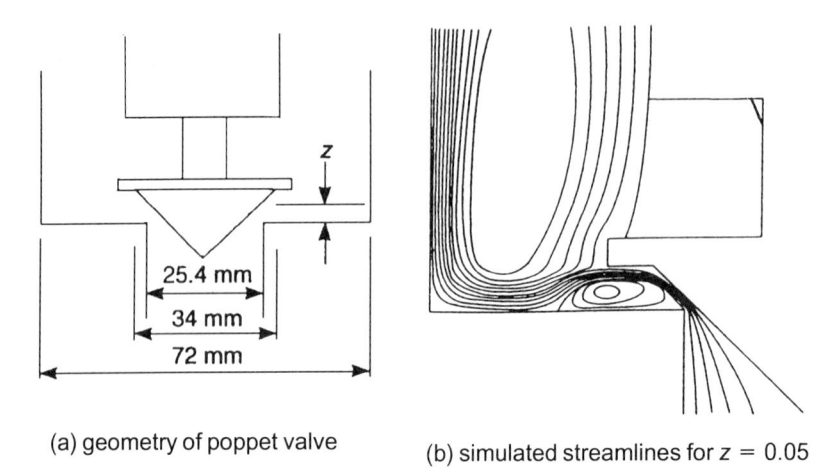

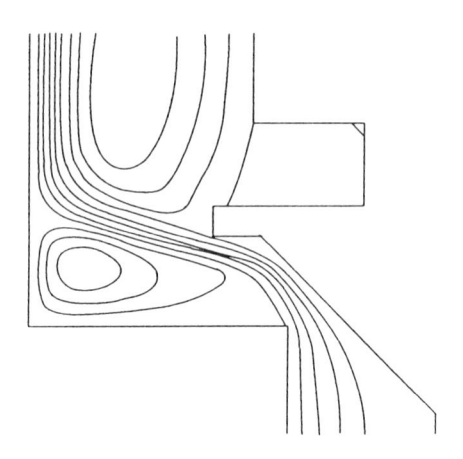

(c) simulated streamlines for z = 0.20

Fig. 11.5 Fluid flow in a poppet valve (reproduced from Vaughan *et al.* 1992 with kind permission)

11.4 FLOW IN OTHER ENGINEERING APPLICATIONS

Vaughan et al. (**50**) used a proprietary finite volume computational fluid dynamics computer program to calculate the flows through poppet valves. They claim that qualitative agreement between simulated and visualized flow patterns was good, but there were errors in the prediction of jet separation and reattachment due to limitations in the finite difference scheme used and in the representation of turbulence. Figure 11.5(a) shows the valve geometry, and Figs. 11.5(b) and (c) show the valve at two different valve openings.

Koutmos and McGuirk (**51**) used a method based on the time-averaged transport equations for momentum, continuity, turbulence, kinetic energy, and energy dissipation to calculate the flow in annular combustor dump diffuser geometries. They reckoned that the loss coefficient calculation was correct to within 7 percent. Figure 11.6(a) shows the geometry. It should be noted that the diffuser is axisymmetric about the centreline of the engine. Figure 11.6(b) shows the direction of flow (stream function contours) in the dump region.

11.5 FLOW IN FLOWMETERS

El Wahed et al. (**52**) report a numerical study of vortex shedding from different shaped bluff bodies, based on a computer code using the stream function-vorticity formulation of the unsteady incompressible flow equations. Figure 11.7(a) shows the flow pattern downstream of a trapezoidal body in a two-dimensional duct and illustrates the periodic nature of the shed vortices. Figure 11.7(b) shows the separation of the shear layer from the sharp leading edges of the body, the strong vortex behind the body and the shed regions of vorticity downstream. A model to allow for turbulence was used in this calculation.

Buckle et al. (**53**) investigated the flow around a floating element or variable area flowmeter as shown in Fig. 11.8(a). The calculations in this work were for laminar flows and compared very well with experimental values, although the rotation of the float was neglected. Figure 11.8(b) which shows one half of the flow, gives an excellent insight into the flow profile development as the flow passes around the float, and indicates the

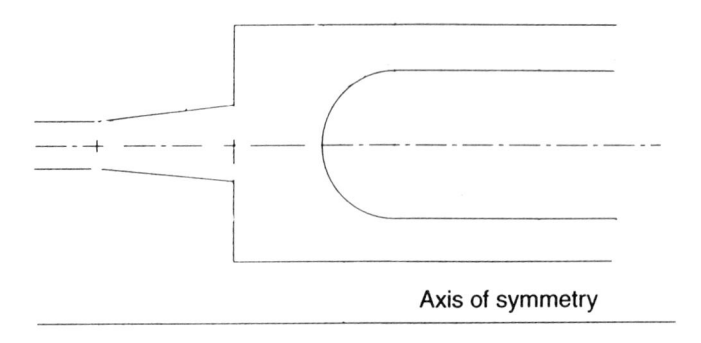

(a) flow configuration investigated

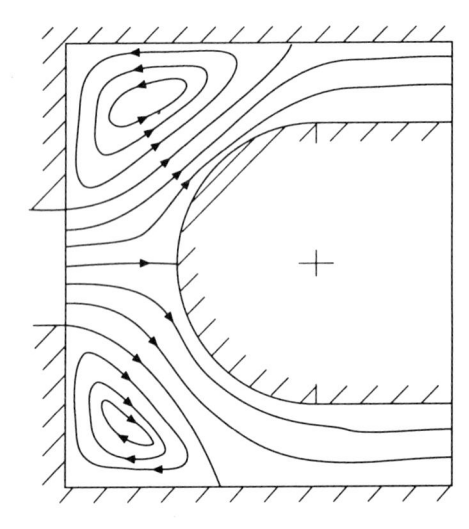

(b) direction of flow (stream function contours)

Fig. 11.6 Flow in an annular dump diffuser (reproduced from Koutmos and McGuirk (51) with kind permission)

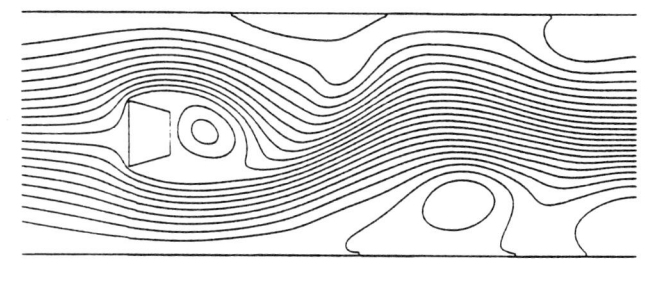

(a) flow past trapezoidal shedder ($Re = 9125$)

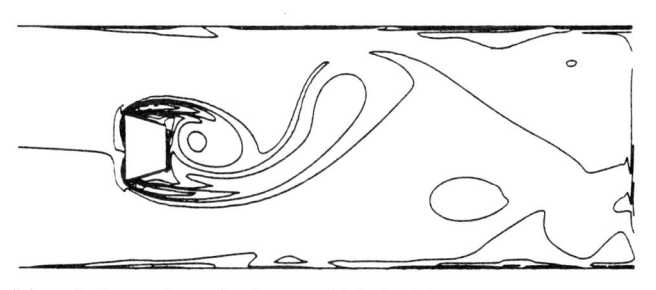

(b) vorticity contours for trapezoidal shedder

Fig. 11.7 Flow through a vortex flowmeter (reproduced from El Wahed et al. (52) with kind permission from Butterworth-Heinemann)

recirculation region after the float. It is this recirculation region which creates the low pressure above the float. This, in turn, contributes to the upward force on the float which balances the weight of the float. The pressure in this region will be close to that in the smallest annulus through which the flow passes. As the float rises, this annulus will increase in size to accommodate the increased flow.

In turbine meters the rotor is generally milled from a solid cylinder and the blades are usually parallel sided with flat ends formed from the ends of the cylinder. The behaviour, therefore, differs from conventional aerofoils or even from flat plates. Xu (**54**) calculated the flow around such turbine flowmeter blades and obtained some very interesting plots of the

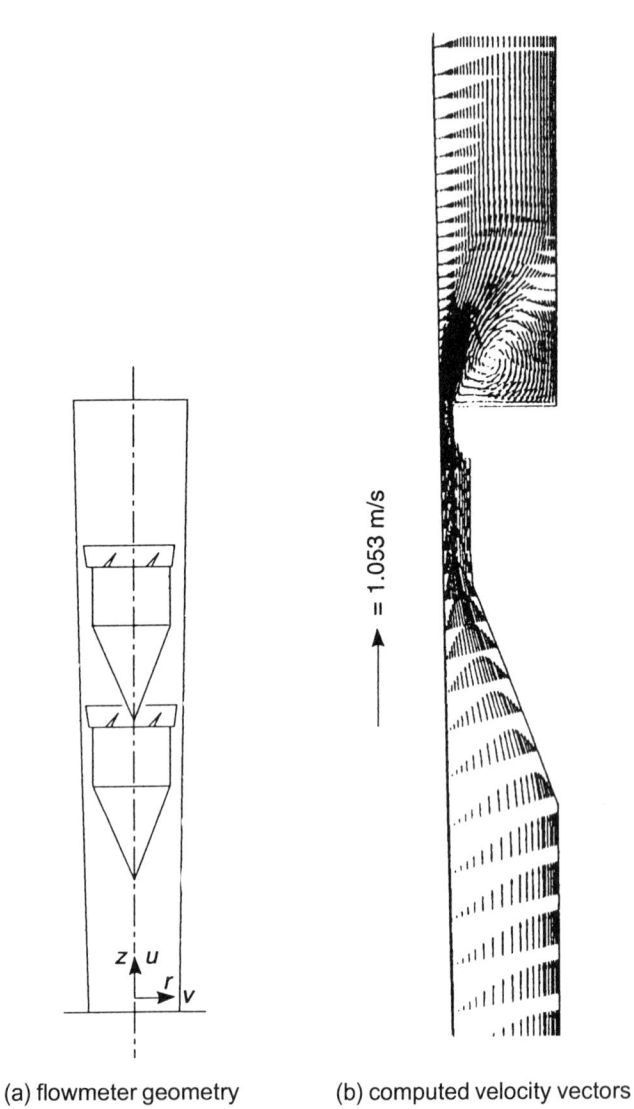

(a) flowmeter geometry (b) computed velocity vectors

Fig. 11.8 Flow through a float in tube meter (reproduced from Buckle et al. (53) with kind permission from Butterworth-Heinemann)

(a) diagram of flow past a flat plate

(b) computed leading edge flow: $-8°$ incidence angle

(c) computed trailing edge flow: $8°$ incidence angle

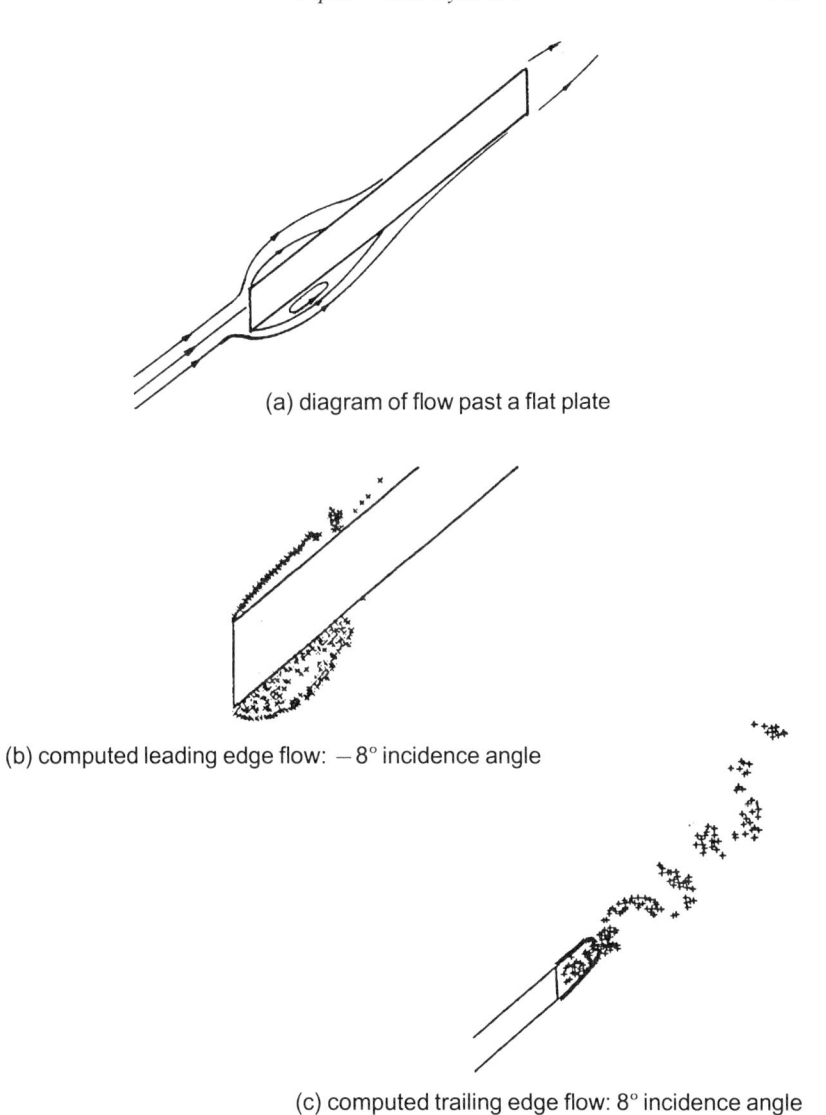

Fig. 11.9 Flow past a flat plate (reproduced from Xu (54) with kind permission from Butterworth-Heinemann)

recirculation regions at the leading edge and at the trailing edge of the blades, some of which are reproduced in Fig. 11.9. Figure 11.9(a) provides a diagram to indicate the overall flow. Figure 11.9(b) shows the recirculation region underneath the plate due to the negative incidence. As the turbine wheel speeds up in response to the flow which it is measuring, the negative incidence will decrease. If there is twist in the blade then the incidence may be small at all radii. However, if the blade is not twisted, it may experience both positive and negative incidence along its length. The trailing edge flow, Fig. 11.9(c), suggests a measure of vortex shedding similar to that encountered in the vortex flowmeter. This is worse than would occur if the plate were an aerofoil section blade. The calibration of such a flowmeter can be adjusted by small changes to the trailing edges resulting from small amounts of filing.

11.6 FUTURE DEVELOPMENTS

The power and accessibility of computational methods is such now that the engineer who is likely to be involved in applications which include the flow of fluids, whether liquids or gases, would be well advised to ensure that he/she has ready access to a user-friendly computer package. There are clearly limitations still to such packages and as the design engineer seeks more and more sophisticated design tools, so these weaknesses will be addressed. Programs for three-dimensional flow clearly devour more computer space and time than two-dimensional programs. Complex geometries may not always be modelled adequately. Turbulent flows are still taxing for the computer, and prediction accuracy is usually limited to the accuracy of the empirical correlations used. In the area of flowmeter flow prediction, the very high precision sought may be difficult to achieve. Each of these qualifications will be gradually overcome, but the user of computer packages should be aware of them and ready to make allowances for them.

There will always be a need for the engineer to have a good grasp of the nature of flows and the simpler solutions to flow problems. Only so can he or she determine the most appropriate course of action and the usefulness of computational methods. It is, therefore, hoped that the topics and treatment of fluid flow in this guide will be found useful.

APPENDIX A
Gas Flow Tables

Tables reproduced with kind permission from Cambridge University Engineering Department. Tables compiled by Dr D. S. Whitehead.

Gas Flow Tables

Specific heat data

Gas	$\gamma = 1.4$ c_p (kJ/kg K)	Gas	$\gamma = 1.333$ c_p (kJ/kg K)
Air	1.006*	CO_2	0.808
N_2	1.041	H_2O (superheated)	1.921†
O_2	0.917	Hot combustion products in gas turbine	1.150
H_2	14.425		
CO	1.042		

*This value of c_p is 5 percent low at 650°C and 15 percent low at 1900°C.

† The approximation that superheated steam behaves as a perfect gas is in general poor, but it does hold reasonably well along an isentropic. The value of c_p given is for use in isentropic calculations. (**55**).

Appendix A: Gas Tables

$\gamma = 1.4$

M	$\dfrac{T}{T_0}$	$\dfrac{p}{p_0}$	$\dfrac{v}{\sqrt{(c_p T_0)}}$	$\dfrac{q_m \sqrt{(c_p T_0)}}{A p_0}$	$\dfrac{F}{q_m \sqrt{(c_p T_0)}}$	$\dfrac{4 f L_{max}}{D}$	$\dfrac{\tfrac{1}{2}\rho v^2}{p_0}$
0.00	1.000	1.000	0.000	0.000	∞	∞	0
0.02	1.000	1.000	0.013	0.044	—	—	0
0.04	1.000	0.999	0.025	0.088	—	—	0.001
0.06	0.999	0.997	0.038	0.133	—	—	0.003
0.08	0.999	0.996	0.051	0.176	—	—	0.005
0.10	0.998	0.993	0.063	0.220	4.576	66.922	0.007
0.12	0.997	0.990	0.076	0.263	3.835	45.408	0.010
0.14	0.996	0.986	0.088	0.306	3.309	32.511	0.014
0.16	0.995	0.982	0.101	0.349	2.917	24.198	0.018
0.18	0.994	0.978	0.113	0.391	2.615	18.543	0.022
0.20	0.992	0.972	0.126	0.432	2.376	14.533	0.027
0.22	0.990	0.967	0.138	0.473	2.182	11.596	0.033
0.24	0.989	0.961	0.151	0.513	2.022	9.386	0.039
0.26	0.987	0.954	0.163	0.553	1.889	7.688	0.045
0.28	0.985	0.947	0.176	0.592	1.777	6.357	0.052
0.30	0.982	0.939	0.188	0.629	1.681	5.299	0.059
0.32	0.980	0.932	0.200	0.667	1.598	4.447	0.067
0.34	0.977	0.923	0.213	0.703	1.526	3.752	0.075
0.36	0.975	0.914	0.225	0.738	1.363	3.180	0.083
0.38	0.972	0.905	0.237	0.772	1.464	2.705	0.091
0.40	0.969	0.896	0.249	0.806	1.361	2.308	0.100
0.42	0.966	0.886	0.261	0.838	1.318	1.974	0.109
0.44	0.963	0.875	0.273	0.869	1.280	1.692	0.119
0.46	0.959	0.865	0.285	0.899	1.247	1.451	0.128
0.48	0.956	0.854	0.297	0.928	1.217	1.245	0.138
0.50	0.952	0.843	0.309	0.956	1.190	1.069	0.148
0.52	0.949	0.832	0.320	0.983	1.167	0.917	0.157
0.54	0.945	0.820	0.332	1.008	1.145	0.787	0.167
0.56	0.941	0.808	0.344	1.033	1.126	0.674	0.177
0.58	0.937	0.796	0.355	1.056	1.109	0.576	0.187
0.60	0.933	0.784	0.367	1.078	1.094	0.491	0.198
0.62	0.929	0.772	0.378	1.099	1.080	0.417	0.208
0.64	0.924	0.759	0.389	1.119	1.068	0.353	0.218
0.66	0.920	0.746	0.400	1.137	1.057	0.298	0.228
0.68	0.915	0.734	0.411	1.154	1.047	0.250	0.238
0.70	0.911	0.721	0.423	1.171	1.038	0.208	0.247
0.72	0.906	0.708	0.433	1.185	1.031	0.172	0.257
0.74	0.901	0.695	0.444	1.199	1.024	0.141	0.266
0.76	0.896	0.682	0.455	1.212	1.018	0.114	0.276
0.78	0.892	0.669	0.466	1.223	1.013	0.092	0.285
0.80	0.887	0.656	0.476	1.234	1.008	0.072	0.294
0.82	0.881	0.643	0.487	1.243	1.004	0.056	0.303
0.84	0.876	0.630	0.497	1.251	1.001	0.042	0.311
0.86	0.871	0.617	0.508	1.259	0.998	0.031	0.319
0.88	0.866	0.604	0.518	1.265	0.996	0.022	0.327
0.90	0.861	0.591	0.528	1.270	0.994	0.015	0.335
0.92	0.855	0.578	0.538	1.274	0.992	0.009	0.343
0.94	0.850	0.566	0.548	1.277	0.991	0.005	0.350
0.96	0.844	0.553	0.558	1.279	0.990	0.002	0.357
0.98	0.839	0.541	0.568	1.281	0.990	0.000	0.363
1.00	0.833	0.528	0.577	1.281	0.990	0.000	0.370

$\gamma = 1.4$

M	$\dfrac{T}{T_0}$	$\dfrac{p}{p_0}$	$\dfrac{v}{\sqrt{(c_p T_0)}}$	$\dfrac{q_m \sqrt{(c_p T_0)}}{A p_0}$	$\dfrac{F}{q_m \sqrt{(c_p T_0)}}$	$\dfrac{4fL_{max}}{D}$	$\dfrac{\tfrac{1}{2}\rho v^2}{p_0}$	M_s	$\dfrac{p_{0s}}{p_0}$	$\dfrac{p_s}{p}$	$\dfrac{p_{0s}}{p}$	$\dfrac{T_s}{T}$	M
1.00	0.833	0.528	0.577	1.281	0.990	0.000	0.370	1.000	1.000	1.000	1.893	1.000	1.00
1.02	0.828	0.516	0.587	1.281	0.990	0.000	0.376	0.981	1.000	1.047	1.938	1.013	1.02
1.04	0.822	0.504	0.596	1.279	0.990	0.002	0.382	0.962	1.000	1.095	1.985	1.026	1.04
1.06	0.817	0.492	0.606	1.277	0.991	0.004	0.387	0.944	1.000	1.144	2.033	1.039	1.06
1.08	0.811	0.480	0.615	1.274	0.992	0.007	0.392	0.928	0.999	1.194	2.082	1.052	1.08
1.10	0.805	0.468	0.624	1.271	0.993	0.010	0.397	0.912	0.999	1.245	2.133	1.065	1.10
1.12	0.799	0.457	0.633	1.267	0.994	0.014	0.401	0.897	0.998	1.297	2.185	1.078	1.12
1.14	0.794	0.445	0.642	1.262	0.995	0.018	0.405	0.882	0.997	1.350	2.239	1.090	1.14
1.16	0.788	0.434	0.651	1.256	0.997	0.023	0.409	0.868	0.996	1.403	2.294	1.103	1.16
1.18	0.782	0.423	0.660	1.250	0.999	0.028	0.413	0.855	0.995	1.458	2.350	1.115	1.18
1.20	0.776	0.412	0.669	1.243	1.000	0.034	0.416	0.842	0.993	1.513	2.408	1.128	1.20
1.22	0.771	0.402	0.677	1.236	1.002	0.039	0.419	0.830	0.991	1.570	2.466	1.141	1.22
1.24	0.765	0.391	0.686	1.228	1.004	0.045	0.421	0.818	0.988	1.627	2.526	1.153	1.24
1.26	0.759	0.381	0.694	1.220	1.007	0.052	0.423	0.807	0.986	1.686	2.588	1.166	1.26
1.28	0.753	0.371	0.703	1.211	1.009	0.058	0.425	0.796	0.983	1.745	2.650	1.178	1.28
1.30	0.747	0.361	0.711	1.201	1.011	0.065	0.427	0.786	0.979	1.805	2.714	1.191	1.30
1.32	0.742	0.351	0.719	1.192	1.014	0.072	0.428	0.776	0.976	1.866	2.778	1.204	1.32
1.34	0.736	0.342	0.727	1.181	1.016	0.079	0.429	0.766	0.972	1.928	2.844	1.216	1.34
1.36	0.730	0.332	0.735	1.171	1.019	0.086	0.430	0.757	0.968	1.991	2.912	1.229	1.36
1.38	0.724	0.323	0.743	1.160	1.021	0.093	0.431	0.748	0.963	2.055	2.980	1.242	1.38
1.40	0.718	0.314	0.750	1.149	1.024	0.100	0.431	0.740	0.958	2.120	3.049	1.255	1.40
1.42	0.713	0.305	0.758	1.138	1.027	0.107	0.431	0.731	0.953	2.186	3.120	1.268	1.42
1.44	0.707	0.297	0.766	1.126	1.029	0.114	0.431	0.723	0.948	2.253	3.192	1.281	1.44
1.46	0.701	0.289	0.773	1.114	1.032	0.121	0.431	0.716	0.942	2.320	3.264	1.294	1.46
1.48	0.695	0.280	0.781	1.102	1.035	0.129	0.430	0.708	0.936	2.389	3.338	1.307	1.48
1.50	0.690	0.272	0.788	1.089	1.038	0.136	0.429	0.701	0.930	2.458	3.413	1.320	1.50

Appendix A: Gas Flow Tables

$\gamma = 1.4$

M	$\dfrac{T}{T_0}$	$\dfrac{p}{p_0}$	$\dfrac{v}{\sqrt{(c_p T_0)}}$	$\dfrac{q_m \sqrt{(c_p T_0)}}{A p_0}$	$\dfrac{F}{q_m \sqrt{(c_p T_0)}}$	$\dfrac{4 f L_{\max}}{D}$	$\dfrac{\tfrac{1}{2}\rho v^2}{p_0}$	M_s	$\dfrac{p_{0s}}{p_0}$	$\dfrac{p_s}{p}$	$\dfrac{p_{0s}}{p}$	$\dfrac{T_s}{T}$	M
1.50	0.690	0.272	0.788	1.089	1.038	0.136	0.429	0.701	0.930	2.458	3.413	1.320	1.50
1.52	0.684	0.265	0.795	1.077	1.041	0.143	0.428	0.694	0.923	2.529	3.489	1.334	1.52
1.54	0.678	0.257	0.802	1.064	1.044	0.151	0.427	0.687	0.917	2.600	3.567	1.347	1.54
1.56	0.673	0.250	0.809	1.051	1.047	0.158	0.425	0.681	0.910	2.673	3.645	1.361	1.56
1.58	0.667	0.242	0.816	1.038	1.050	0.165	0.423	0.675	0.903	2.746	3.725	1.374	1.58
1.60	0.661	0.235	0.823	1.025	1.053	0.172	0.422	0.668	0.895	2.820	3.805	1.388	1.60
1.62	0.656	0.228	0.830	1.011	1.056	0.180	0.420	0.663	0.888	2.895	3.887	1.402	1.62
1.64	0.650	0.222	0.836	0.998	1.059	0.187	0.417	0.657	0.880	2.971	3.969	1.416	1.64
1.66	0.645	0.215	0.843	0.985	1.061	0.194	0.415	0.651	0.872	3.048	4.053	1.430	1.66
1.68	0.639	0.209	0.849	0.971	1.064	0.201	0.413	0.646	0.864	3.126	4.138	1.444	1.68
1.70	0.634	0.203	0.856	0.958	1.067	0.208	0.410	0.641	0.856	3.205	4.224	1.458	1.70
1.72	0.628	0.197	0.862	0.944	1.070	0.215	0.407	0.635	0.847	3.285	4.311	1.473	1.72
1.74	0.623	0.191	0.869	0.931	1.073	0.222	0.404	0.631	0.839	3.366	4.399	1.487	1.74
1.76	0.617	0.185	0.875	0.917	1.076	0.228	0.401	0.626	0.830	3.447	4.488	1.502	1.76
1.78	0.612	0.179	0.881	0.904	1.079	0.235	0.398	0.621	0.822	3.530	4.578	1.517	1.78
1.80	0.607	0.174	0.887	0.890	1.082	0.242	0.395	0.617	0.813	3.613	4.670	1.532	1.80
1.82	0.602	0.169	0.893	0.877	1.085	0.249	0.391	0.612	0.804	3.698	4.762	1.547	1.82
1.84	0.596	0.164	0.899	0.863	1.088	0.255	0.388	0.608	0.795	3.783	4.855	1.562	1.84
1.86	0.591	0.159	0.904	0.850	1.091	0.262	0.384	0.604	0.786	3.870	4.950	1.577	1.86
1.88	0.586	0.154	0.910	0.837	1.094	0.268	0.381	0.600	0.777	3.957	5.045	1.592	1.88
1.90	0.581	0.149	0.916	0.824	1.097	0.274	0.377	0.596	0.767	4.045	5.142	1.608	1.90
1.92	0.576	0.145	0.921	0.811	1.100	0.281	0.373	0.592	0.758	4.134	5.239	1.624	1.92
1.94	0.571	0.140	0.927	0.798	1.103	0.287	0.370	0.588	0.749	4.224	5.338	1.639	1.94
1.96	0.566	0.136	0.932	0.785	1.106	0.293	0.366	0.584	0.740	4.315	5.438	1.655	1.96
1.98	0.561	0.132	0.938	0.772	1.108	0.299	0.362	0.581	0.730	4.407	5.539	1.671	1.98
2.00	0.556	0.128	0.943	0.759	1.111	0.305	0.358	0.577	0.721	4.500	5.641	1.688	2.00

$\gamma = 1.4$

M	$\dfrac{T}{T_0}$	$\dfrac{p}{p_0}$	$\dfrac{v}{\sqrt{(c_p T_0)}}$	$\dfrac{q_m \sqrt{(c_p T_0)}}{A p_0}$	$\dfrac{F}{q_m \sqrt{(c_p T_0)}}$	$\dfrac{4fL_{max}}{D}$	$\dfrac{\frac{1}{2}\rho v^2}{p_0}$	M_s	$\dfrac{p_{0s}}{p_0}$	$\dfrac{p_s}{p}$	$\dfrac{p_{0s}}{p}$	$\dfrac{T_s}{T}$	M
2.00	0.556	0.1278	0.943	0.759	1.111	0.305	0.358	0.577	0.721	4.50	5.64	1.688	2.00
2.02	0.551	0.1239	0.948	0.747	1.114	0.311	0.354	0.574	0.712	4.59	5.74	1.704	2.02
2.04	0.546	0.1201	0.953	0.743	1.117	0.317	0.350	0.571	0.702	4.69	5.85	1.720	2.04
2.06	0.541	0.1164	0.958	0.722	1.119	0.323	0.346	0.567	0.693	4.78	5.95	1.737	2.06
2.08	0.536	0.1128	0.963	0.709	1.122	0.328	0.342	0.564	0.684	4.88	6.06	1.754	2.08
2.10	0.531	0.1094	0.968	0.697	1.125	0.334	0.338	0.561	0.674	4.98	6.17	1.770	2.10
2.12	0.527	0.1060	0.973	0.685	1.128	0.339	0.333	0.558	0.665	5.08	6.27	1.787	2.12
2.14	0.522	0.1027	0.978	0.674	1.130	0.345	0.329	0.555	0.656	5.18	6.38	1.805	2.14
2.16	0.517	0.0996	0.983	0.662	1.133	0.350	0.325	0.553	0.646	5.28	6.49	1.822	2.16
2.18	0.513	0.0965	0.987	0.650	1.136	0.356	0.321	0.550	0.637	5.38	6.60	1.839	2.18
2.20	0.508	0.0935	0.992	0.639	1.138	0.361	0.317	0.547	0.628	5.48	6.72	1.857	2.20
2.22	0.504	0.0906	0.996	0.628	1.141	0.366	0.313	0.544	0.619	5.58	6.83	1.875	2.22
2.24	0.499	0.0878	1.001	0.617	1.143	0.371	0.309	0.542	0.610	5.69	6.94	1.892	2.24
2.26	0.495	0.0851	1.005	0.606	1.146	0.376	0.304	0.539	0.601	5.79	7.06	1.910	2.26
2.28	0.490	0.0825	1.010	0.595	1.148	0.381	0.300	0.537	0.592	5.90	7.18	1.929	2.28
2.30	0.486	0.0800	1.014	0.584	1.151	0.386	0.296	0.534	0.583	6.01	7.29	1.947	2.30
2.32	0.482	0.0775	1.018	0.574	1.153	0.391	0.292	0.532	0.575	6.11	7.41	1.965	2.32
2.34	0.477	0.0751	1.022	0.563	1.156	0.396	0.288	0.530	0.566	6.22	7.53	1.984	2.34
2.36	0.473	0.0728	1.027	0.553	1.158	0.401	0.284	0.527	0.557	6.33	7.65	2.003	2.36
2.38	0.469	0.0706	1.031	0.543	1.161	0.405	0.280	0.525	0.549	6.44	7.77	2.021	2.38
2.40	0.465	0.0684	1.035	0.533	1.163	0.410	0.276	0.523	0.540	6.55	7.90	2.040	2.40
2.42	0.461	0.0663	1.039	0.523	1.165	0.414	0.272	0.521	0.532	6.67	8.02	2.060	2.42
2.44	0.456	0.0643	1.043	0.514	1.168	0.419	0.268	0.519	0.523	6.78	8.15	2.079	2.44
2.46	0.452	0.0623	1.046	0.504	1.170	0.423	0.264	0.517	0.515	6.89	8.27	2.098	2.46
2.48	0.448	0.0604	1.050	0.495	1.172	0.428	0.260	0.515	0.507	7.01	8.40	2.118	2.48
2.50	0.444	0.0585	1.054	0.486	1.175	0.432	0.256	0.513	0.499	7.13	8.53	2.138	2.50

Appendix A: Gas Flow Tables

$\gamma = 1.4$

M	$\dfrac{T}{T_0}$	$\dfrac{p}{p_0}$	$\dfrac{v}{\sqrt{(c_p T_0)}}$	$\dfrac{q_m \sqrt{(c_p T_0)}}{A p_0}$	$\dfrac{F}{q_m \sqrt{(c_p T_0)}}$	$\dfrac{4 f L_{\max}}{D}$	$\dfrac{\tfrac{1}{2}\rho v^2}{p_0}$	M_s	$\dfrac{p_{0s}}{p_0}$	$\dfrac{p_s}{p}$	$\dfrac{p_{0s}}{p}$	$\dfrac{T_s}{T}$	M
2.5	0.444	0.0585	1.054	0.486	1.175	0.432	0.256	0.513	0.499	7.13	8.53	2.14	2.5
2.6	0.425	0.0501	1.072	0.442	1.186	0.453	0.237	0.504	0.460	7.72	9.18	2.24	2.6
2.7	0.407	0.0430	1.089	0.402	1.196	0.472	0.219	0.496	0.424	8.34	9.86	2.34	2.7
2.8	0.389	0.0368	1.105	0.366	1.206	0.490	0.202	0.488	0.389	8.98	10.57	2.45	2.8
2.9	0.373	0.0317	1.120	0.333	1.215	0.507	0.186	0.481	0.358	9.65	11.30	2.56	2.9
3.0	0.357	0.0272	1.134	0.303	1.224	0.522	0.172	0.475	0.328	10.33	12.06	2.68	3.0
3.1	0.342	0.0234	1.147	0.275	1.232	0.537	0.158	0.470	0.301	11.05	12.85	2.80	3.1
3.2	0.328	0.0202	1.159	0.250	1.240	0.550	0.145	0.464	0.276	11.78	13.65	2.92	3.2
3.3	0.315	0.0175	1.171	0.228	1.248	0.563	0.133	0.460	0.253	12.54	14.5	3.05	3.3
3.4	0.302	0.0151	1.182	0.207	1.255	0.575	0.122	0.455	0.232	13.32	15.4	3.18	3.4
3.5	0.290	0.0131	1.192	0.189	1.261	0.586	0.112	0.451	0.213	14.13	16.2	3.32	3.5
3.6	0.278	0.0114	1.201	0.172	1.268	0.597	0.103	0.447	0.195	14.95	17.2	3.45	3.6
3.7	0.268	0.0099	1.210	0.157	1.274	0.607	0.0049	0.444	0.179	15.80	18.1	3.60	3.7
3.8	0.257	0.0086	1.219	0.143	1.279	0.616	0.0872	0.441	0.164	16.68	19.1	3.74	3.8
3.9	0.247	0.0075	1.227	0.131	1.284	0.625	0.0802	0.438	0.151	17.58	20.1	3.89	3.9
4.0	0.238	6.59×10^{-3}	1.234	1.20×10^{-1}	1.290	0.633	7.38×10^{-2}	0.435	1.39×10^{-1}	18.50	21.1	4.05	4.0
4.5	0.198	3.45×10^{-3}	1.266	7.73×10^{-2}	1.311	0.668	4.90×10^{-2}	0.424	9.17×10^{-2}	23.46	26.5	4.87	4.5
5.0	0.167	1.89×10^{-3}	1.291	5.12×10^{-2}	1.328	0.694	3.31×10^{-2}	0.415	6.17×10^{-2}	29.00	32.7	5.80	5.0
5.5	0.142	1.07×10^{-3}	1.310	3.47×10^{-2}	1.341	0.714	2.26×10^{-2}	0.409	4.22×10^{-2}	35.13	39.4	6.83	5.5
6.0	0.122	6.33×10^{-4}	1.325	2.41×10^{-2}	1.351	0.730	1.60×10^{-2}	0.404	2.97×10^{-2}	41.83	46.8	7.94	6.0
6.5	0.106	3.85×10^{-4}	1.337	1.70×10^{-2}	1.360	0.743	1.14×10^{-2}	0.400	2.11×10^{-2}	49.13	54.9	9.13	6.5
7.0	0.093	2.42×10^{-4}	1.347	1.23×10^{-2}	1.367	0.753	8.29×10^{-3}	0.397	1.54×10^{-2}	57.00	63.6	10.5	7.0
7.5	0.082	1.55×10^{-4}	1.355	9.03×10^{-3}	1.372	0.761	6.08×10^{-3}	0.395	1.13×10^{-2}	65.46	72.9	11.8	7.5
8.0	0.072	1.02×10^{-4}	1.362	6.74×10^{-3}	1.377	0.768	4.59×10^{-3}	0.393	8.49×10^{-3}	74.50	82.9	13.4	8.0
8.5	0.065	6.90×10^{-5}	1.368	5.10×10^{-3}	1.381	0.774	3.48×10^{-3}	0.391	6.45×10^{-3}	84.13	93.5	14.9	8.5
9.0	0.058	4.74×10^{-5}	1.372	3.92×10^{-3}	1.385	0.779	2.69×10^{-3}	0.390	4.96×10^{-3}	94.33	104.8	16.7	9.0
9.5	0.052	3.31×10^{-5}	1.377	3.04×10^{-3}	1.387	0.783	2.11×10^{-3}	0.389	3.86×10^{-3}	105.1	116.7	18.8	9.5
∞	0	0	1.414	0	1.414	0.822	0	0.378	0	∞	∞	∞	∞

$\gamma = 1.333$

M	$\dfrac{T}{T_0}$	$\dfrac{p}{p_0}$	$\dfrac{v}{\sqrt{(c_v T_0)}}$	$\dfrac{q_m \sqrt{(c_v T_0)}}{A p_0}$	$\dfrac{F}{q_m \sqrt{(c_v T_0)}}$	$\dfrac{4 f L_{\max}}{D}$
0.000	1.000	1.000	0.000	0.000	∞	∞
0.020	1.000	1.000	0.012	0.046	—	—
0.040	1.000	0.999	0.023	0.092	—	—
0.060	0.999	0.998	0.035	0.138	—	—
0.080	0.999	0.996	0.046	0.184	—	—
0.100	0.998	0.993	0.058	0.230	4.383	70.372
0.120	0.998	0.990	0.069	0.275	3.672	47.768
0.140	0.997	0.987	0.081	0.320	3.168	34.216
0.160	0.996	0.983	0.092	0.364	2.792	25.478
0.180	0.995	0.979	0.104	0.408	2.502	19.533
0.200	0.993	0.974	0.115	0.451	2.272	15.317
0.220	0.992	0.968	0.126	0.494	2.086	12.227
0.240	0.991	0.963	0.138	0.536	1.933	9.903
0.260	0.989	0.956	0.149	0.578	1.805	8.115
0.280	0.987	0.949	0.161	0.618	1.697	6.714
0.300	0.985	0.942	0.172	0.658	1.604	5.600
0.320	0.983	0.935	0.183	0.697	1.525	4.702
0.340	0.981	0.927	0.194	0.735	1.456	3.969
0.360	0.979	0.918	0.206	0.772	1.395	3.366
0.380	0.977	0.909	0.217	0.808	1.342	2.866
0.400	0.974	0.900	0.228	0.843	1.296	2.447
0.420	0.971	0.891	0.239	0.877	1.255	2.094
0.440	0.969	0.881	0.250	0.909	1.218	1.795
0.460	0.966	0.871	0.261	0.941	1.186	1.541
0.480	0.963	0.860	0.272	0.972	1.157	1.323
0.500	0.960	0.849	0.283	1.001	1.131	1.137
0.520	0.957	0.838	0.294	1.029	1.108	0.976
0.540	0.954	0.827	0.304	1.057	1.087	0.837
0.560	0.950	0.816	0.315	1.082	1.069	0.717
0.580	0.947	0.804	0.326	1.107	1.052	0.614
0.600	0.943	0.792	0.336	1.130	1.037	0.523
0.620	0.940	0.780	0.347	1.152	1.024	0.445
0.640	0.936	0.768	0.357	1.173	1.012	0.377
0.660	0.932	0.756	0.368	1.193	1.001	0.318
0.680	0.929	0.743	0.378	1.211	0.992	0.267
0.700	0.925	0.731	0.388	1.229	0.983	0.223
0.720	0.921	0.718	0.399	1.244	0.976	0.184
0.740	0.916	0.705	0.409	1.259	0.969	0.151
0.760	0.912	0.692	0.419	1.273	0.963	0.123
0.780	0.908	0.680	0.429	1.285	0.958	0.098
0.800	0.904	0.667	0.439	1.296	0.953	0.078
0.820	0.899	0.654	0.449	1.306	0.949	0.060
0.840	0.895	0.641	0.459	1.315	0.946	0.045
0.860	0.890	0.628	0.468	1.323	0.943	0.033
0.880	0.886	0.615	0.478	1.329	0.941	0.023
0.900	0.881	0.603	0.488	1.335	0.939	0.016
0.920	0.876	0.590	0.497	1.339	0.938	0.010
0.940	0.872	0.577	0.506	1.343	0.936	0.005
0.960	0.867	0.565	0.516	1.345	0.936	0.002
0.980	0.862	0.552	0.525	1.346	0.935	0.001
1.000	0.857	0.540	0.534	1.347	0.935	0.000

Appendix A: Gas Flow Tables

$\gamma = 1.333$

M	$\dfrac{T}{T_0}$	$\dfrac{p}{p_0}$	$\dfrac{v}{\sqrt{(c_v T_0)}}$	$\dfrac{q_m \sqrt{(c_v T_0)}}{A p_0}$	$\dfrac{F}{q_m \sqrt{(c_v T_0)}}$	$\dfrac{4 f L_{\max}}{D}$	M_s
1.000	0.857	0.540	0.534	1.347	0.935	0.000	1.000
1.020	0.852	0.528	0.543	1.346	0.935	0.000	0.981
1.040	0.847	0.515	0.552	1.345	0.936	0.002	0.962
1.060	0.842	0.503	0.561	1.343	0.936	0.004	0.944
1.080	0.837	0.491	0.570	1.340	0.937	0.007	0.927
1.100	0.832	0.480	0.579	1.336	0.938	0.011	0.911
1.120	0.827	0.468	0.588	1.331	0.939	0.015	0.896
1.140	0.822	0.457	0.596	1.326	0.941	0.020	0.881
1.160	0.817	0.445	0.605	1.320	0.942	0.025	0.867
1.180	0.812	0.434	0.614	1.313	0.944	0.031	0.854
1.200	0.807	0.423	0.622	1.306	0.946	0.037	0.841
1.220	0.801	0.412	0.630	1.298	0.948	0.043	0.829
1.240	0.796	0.402	0.638	1.289	0.950	0.049	0.817
1.260	0.791	0.391	0.647	1.280	0.952	0.056	0.805
1.280	0.786	0.381	0.655	1.270	0.954	0.063	0.794
1.300	0.780	0.371	0.663	1.260	0.957	0.071	0.784
1.320	0.775	0.361	0.671	1.249	0.959	0.078	0.774
1.340	0.770	0.351	0.678	1.238	0.962	0.086	0.764
1.360	0.765	0.341	0.686	1.227	0.965	0.093	0.754
1.380	0.759	0.332	0.694	1.215	0.967	0.101	0.745
1.400	0.754	0.323	0.701	1.202	0.970	0.109	0.736
1.420	0.749	0.314	0.709	1.190	0.973	0.117	0.728
1.440	0.743	0.305	0.716	1.177	0.976	0.125	0.720
1.460	0.738	0.296	0.724	1.164	0.979	0.133	0.712
1.480	0.733	0.288	0.731	1.150	0.981	0.141	0.704
1.500	0.727	0.280	0.738	1.137	0.984	0.149	0.697
1.520	0.722	0.272	0.745	1.123	0.987	0.157	0.689
1.540	0.717	0.264	0.752	1.109	0.990	0.165	0.682
1.560	0.712	0.256	0.759	1.094	0.994	0.173	0.676
1.580	0.706	0.249	0.766	1.080	0.997	0.181	0.669
1.600	0.701	0.241	0.773	1.066	1.000	0.190	0.663
1.620	0.696	0.234	0.780	1.051	1.003	0.198	0.657
1.640	0.691	0.227	0.787	1.036	1.006	0.205	0.651
1.660	0.685	0.221	0.793	1.021	1.009	0.213	0.645
1.680	0.680	0.214	0.800	1.007	1.012	0.221	0.639
1.700	0.675	0.208	0.806	0.992	1.015	0.229	0.634
1.720	0.670	0.201	0.812	0.977	1.018	0.237	0.629
1.740	0.665	0.195	0.819	0.962	1.022	0.245	0.623
1.760	0.660	0.189	0.825	0.947	1.025	0.252	0.618
1.780	0.655	0.183	0.831	0.932	1.028	0.260	0.614
1.800	0.650	0.178	0.837	0.917	1.031	0.267	0.609
1.820	0.645	0.172	0.843	0.903	1.034	0.275	0.604
1.840	0.640	0.167	0.849	0.888	1.037	0.282	0.600
1.860	0.635	0.162	0.855	0.873	1.040	0.289	0.595
1.880	0.630	0.157	0.861	0.858	1.043	0.297	0.591
1.900	0.625	0.152	0.867	0.844	1.047	0.304	0.587
1.920	0.620	0.147	0.872	0.830	1.050	0.311	0.583
1.940	0.615	0.143	0.878	0.815	1.053	0.318	0.579
1.960	0.610	0.138	0.883	0.801	1.056	0.325	0.575
1.980	0.605	0.134	0.889	0.787	1.059	0.332	0.572
2.000	0.600	0.130	0.894	0.773	1.062	0.339	0.568

$\gamma = 1.333$

M	$\dfrac{T}{T_0}$	$\dfrac{p}{p_0}$	$\dfrac{v}{\sqrt{(c_v T_0)}}$	$\dfrac{q_m \sqrt{(c_v T_0)}}{A p_0}$	$\dfrac{F}{q_m \sqrt{(c_v T_0)}}$	$\dfrac{4 f L_{\max}}{D}$	M_s
2.000	0.600	0.1296	0.894	0.773	1.062	0.339	0.568
2.020	0.595	0.1255	0.899	0.759	1.065	0.345	0.564
2.040	0.591	0.1216	0.905	0.745	1.068	0.352	0.561
2.060	0.586	0.1177	0.910	0.732	1.071	0.359	0.558
2.080	0.581	0.1140	0.915	0.718	1.074	0.365	0.554
2.100	0.577	0.1104	0.920	0.705	1.077	0.371	0.551
2.120	0.572	0.1069	0.925	0.692	1.080	0.378	0.548
2.140	0.567	0.1035	0.930	0.679	1.083	0.384	0.545
2.160	0.563	0.1002	0.935	0.696	1.085	0.390	0.542
2.180	0.558	0.0970	0.940	0.653	1.088	0.396	0.539
2.200	0.554	0.0939	0.945	0.641	1.091	0.402	0.536
2.220	0.549	0.0909	0.949	0.629	1.094	0.408	0.533
2.240	0.545	0.0880	0.954	0.617	1.097	0.414	0.530
2.260	0.540	0.0851	0.959	0.605	1.100	0.420	0.528
2.280	0.536	0.0824	0.963	0.593	1.102	0.426	0.525
2.300	0.532	0.0798	0.968	0.581	1.105	0.431	0.523
2.320	0.527	0.0772	0.972	0.570	1.108	0.437	0.520
2.340	0.523	0.0747	0.977	0.558	1.110	0.442	0.518
2.360	0.519	0.0723	0.981	0.547	1.113	0.448	0.515
2.380	0.515	0.0700	0.985	0.536	1.116	0.453	0.513
2.400	0.510	0.0678	0.989	0.526	1.118	0.459	0.511
2.420	0.506	0.0656	0.994	0.515	1.121	0.464	0.508
2.440	0.502	0.0635	0.998	0.505	1.124	0.469	0.506
2.460	0.498	0.0614	1.002	0.495	1.126	0.474	0.504
2.480	0.494	0.0595	1.006	0.485	1.129	0.479	0.502
2.500	0.490	0.0575	1.010	0.475	1.131	0.484	0.500
2.600	0.470	0.0489	1.029	0.428	1.143	0.508	0.490
2.700	0.452	0.0415	1.047	0.385	1.155	0.530	0.481
2.800	0.434	0.0353	1.064	0.347	1.166	0.551	0.473
2.900	0.417	0.0300	1.080	0.312	1.177	0.571	0.466
3.000	0.400	0.0256	1.095	0.280	1.187	0.589	0.460
3.100	0.385	0.0218	1.109	0.252	1.196	0.607	0.453
3.200	0.370	0.0186	1.123	0.226	1.205	0.623	0.448
3.300	0.355	0.0159	1.135	0.204	1.214	0.638	0.443
3.400	0.342	0.0136	1.147	0.183	1.222	0.652	0.438
3.500	0.329	0.0117	1.158	0.165	1.229	0.665	0.434
3.600	0.317	0.0100	1.169	0.148	1.237	0.678	0.430
3.700	0.305	0.0086	1.179	0.133	1.244	0.690	0.426
3.800	0.294	0.0074	1.188	0.120	1.250	0.701	0.422
3.900	0.283	0.0064	1.197	0.108	1.256	0.712	0.419
4.000	0.273	0.0055	1.206	0.098	1.262	0.721	0.416

APPENDIX B
Steam Tables

Taken from *Thermodynamic tables in SI (metric) units* by R. W. Hayward (**47**) and reproduced with kind permission of Cambridge University Press.

Density of water and steam

[0.1 MN/m² = 1 bar ≈ 14.5 lbf/in²]

Density/(kg/m³)

Pressure/(MN/m²)	0	0	0.01	0.05	0.1	0.5	1	2	4	6	8	10	15	20	22.12	25	30	40	50	100	Celsius temp., °C
Pressure/bar	0	0	0.1	0.5	1	5	10	20	40	60	80	100	150	200	221.2	250	300	400	500	1000	
Sat. Celsius temp., °C	—	—	45.8	81.3	99.6	151.8	179.9	212.4	250.3	275.6	295.0	311.0	342.1	365.7	374.15	—	—	—	—	—	
Sat. density {Water	—	—	990	971	958	915	887	850	799	758	722	688	603	491	315	—	—	—	—	—	
kg/m³ {Steam	—	—	0.0681	0.309	0.590	2.67	5.15	10.05	20.10	30.8	42.5	55.4	96.7	170.2	315	—	—	—	—	—	
Celsius temp., °C																					
50	0.0672		988	988	988	988	988	989	990	991	992	992	994	997	997	999	1001	1005	1009	1027	50
75	0.0624		975	975	975	975	975	976	976	977	978	979	981	984	984	986	988	992	996	1015	75
100	0.0582	0.293	0.590	958	958	959	959	959	960	961	962	963	965	967	968	970	972	976	980	1000	100
125	0.0545	0.274	0.550	939	939	939	940	941	942	943	944	946	949	950	951	954	958	963	984		125
150	0.0512	0.257	0.516	917	918	918	919	920	921	922	925	928	929	930	933	938	943	965			150
175	0.0484	0.243	0.487	2.504	892	893	894	896	897	898	901	904	906	907	916	921	946				175
200	0.0458	0.230	0.460	2.353	4.86	865	867	868	870	871	875	878	880	882	885	891	897	924			200
225	0.0435	0.218	0.437	2.223	4.55	9.64	835	837	839	841	845	849	851	853	857	864	871	901			225
250	0.0414	0.207	0.416	2.108	4.30	8.97	799	802	804	806	811	817	819	821	826	835	843	877			250
275	0.0395	0.198	0.396	2.006	4.07	8.43	18.33	759	762	765	773	779	782	785	791	802	811	850			275
300	0.0378	0.189	0.379	1.914	3.88	7.97	17.00	27.67	41.2	715	726	735	739	743	751	765	777	823			300
325	0.0362	0.181	0.363	1.830	3.70	7.57	15.93	25.41	36.5	50.4	665	680	685	692	703	722	738	793			325
350	0.0348	0.174	0.348	1.754	3.54	7.22	15.05	23.68	33.4	44.6	87.2	600	612	625	644	671	693	761			350
375	0.0334	0.167	0.335	1.685	3.40	6.90	14.29	22.28	31.0	40.8	72.0	130.5	218.4	504	558	609	641	727			375
400	0.0322	0.161	0.322	1.620	3.26	6.62	13.63	21.11	29.1	37.9	63.9	100.5	122.6	166.3	353.3	523.8	578.2	691.4			400
425	0.0310	0.155	0.311	1.561	3.14	6.36	13.04	20.09	27.6	35.6	58.4	87.2	102.2	126.8	188.3	392.4	498.3	654.1			425
450	0.0300	0.150	0.300	1.506	3.03	6.12	12.51	19.19	26.2	33.6	54.2	78.7	90.7	109.0	148.5	272.1	401.3	613.9			450
475	0.0290	0.145	0.290	1.455	2.92	5.99	12.02	18.39	25.0	32.0	50.9	72.5	82.8	97.9	128.3	210.3	316.0	571.4			475
500	0.0280	0.140	0.280	1.407	2.83	5.70	11.58	17.67	24.0	30.5	48.1	67.7	76.8	89.9	115.2	178.1	257.6	528.2			500
550	0.0263	0.132	0.263	1.320	2.65	5.33	10.80	16.41	22.2	28.1	43.7	60.4	68.0	78.6	98.4	143.2	195.6	445.3			550
600	0.0248	0.124	0.248	1.244	2.49	5.01	10.13	15.34	20.7	26.1	40.2	55.1	61.6	70.8	87.4	123.6	163.6	374.8			600
650	0.0235	0.117	0.235	1.176	2.36	4.73	9.54	14.42	19.4	24.4	37.4	50.8	56.7	64.9	79.5	110.5	143.7	322.0			650
700	0.0223	0.111	0.223	1.115	2.23	4.48	9.02	13.61	18.3	23.0	35.0	47.4	52.7	60.1	73.3	100.7	129.5	282.8			700
750	0.0212	0.106	0.212	1.060	2.12	4.26	8.55	12.89	17.3	21.7	32.9	44.4	49.4	56.2	68.2	93.0	118.8	253.0			750
800	0.0202	0.101	0.202	1.011	2.02	4.05	8.14	12.26	16.4	20.6	31.2	41.9	46.6	52.9	64.0	86.8	110.2	230.4			800

Critical isobar

Note: **Density** is tabulated here, instead of specific volume, since interpolation between pressures is thereby facilitated.

APPENDIX C
Steam Chart

SPECIFIC ENTROPY (J/kg K)

APPENDIX D

Boundary Layer Equations

General equations

In Chapter 5 the importance of the boundary layer was briefly reviewed. In this appendix the equations of flow in two dimensions are given and from these the boundary layer approximation is derived. (This treatment has benefited from various authors **(19) (56) (57)** and colleagues.) Using V_x for the velocity in the x direction parallel with the main stream, or the flow boundary, and V_y for the flow perpendicular to this, the equations of motion may be set down as

$$\underbrace{\underbrace{\frac{\partial V_x}{\partial t}}_{\text{linear}} + \underbrace{V_x \frac{\partial V_x}{\partial x} + V_y \frac{\partial V_x}{\partial y}}_{\text{Squared}}}_{\text{Inertia terms}} = \overbrace{-\frac{1}{\rho}\frac{\partial p}{\partial x}}^{\text{Pressure gradient terms}} + \overbrace{v\left(\frac{\partial^2 V_x}{\partial x^2} + \frac{\partial^2 V_x}{\partial y^2}\right)}^{\text{Viscous terms}} \qquad (D1)$$

$$\frac{\partial V_y}{\partial t} + V_x \frac{\partial V_y}{\partial x} + V_y \frac{\partial V_y}{\partial y} = -\frac{1}{\rho}\frac{\partial p}{\partial y} + v\left(\frac{\partial^2 V_y}{\partial x^2} + \frac{\partial^2 V_y}{\partial y^2}\right) \qquad (D2)$$

The continuity equation is

$$\frac{\partial V_x}{\partial x} + \frac{\partial V_y}{\partial y} = 0 \qquad (D3)$$

These equations are non-linear. That is to say that terms appear which have velocity squared. There are three traditional methods of solving these equations (in the absence of computer methods):

(i) by seeking situations for which the squared terms are zero;
(ii) by seeking situations for which the inertia terms are negligible. This tends to be at very low Reynolds numbers;
(iii) by seeking situations for which the viscous terms are negligible. This is known as inviscid theory.

Laminar boundary layer equations

The first step is to simplify the equations so that they are reduced to a manageable form. To do this the approximate magnitudes of the terms are considered using Fig. D1.

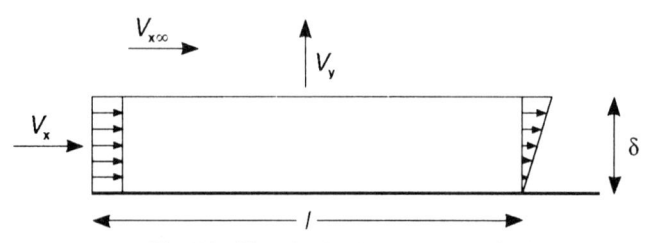

Fig. D1 Magnitudes based on continuity

From Fig. 5.3 it was noted that since more fluid enters the boundary layer parallel to the boundary wall than leaves it, there must be a velocity component perpendicular to the wall and away from it. Figure D1 shows that for a uniform profile at the leading edge of a plate and a 'linear' boundary layer at l downstream, the net inflow parallel to the wall is equal to the net outflow perpendicular to the wall and hence

$$V_y l \approx \left(V_{x\infty} - \frac{V_{x\infty}}{2} \right) \delta$$

$$\therefore \quad V_y \approx \frac{V_{x\infty}}{2l} \delta$$

This indicates approximately that the velocity perpendicular to the wall to the velocity parallel to the wall is in the ratio of the boundary layer thickness to the distance from the leading edge of the plate. If a similar argument is used for the momentum balance shown in Fig. D2, an approximation relating the viscous force on the plate to the change in

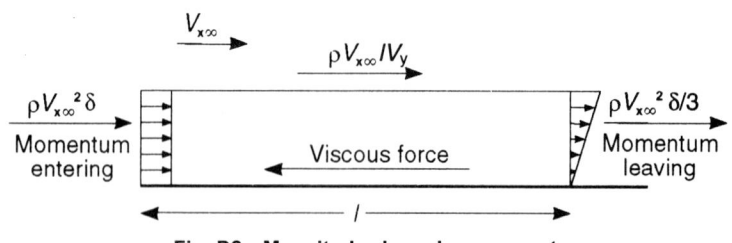

Fig. D2 Magnitudes based on momentum

Appendix D: Boundary Layer Equations

momentum parallel to the plate is obtained

$$\frac{\mu V_{x\infty} l}{\delta} \approx \rho V_{x\infty}^2 \delta - \frac{\rho V_{x\infty}^2 \delta}{3} - \rho V_{x\infty} l V_y$$

$$\frac{\mu V_{x\infty} l}{\delta} \approx \rho V_{x\infty}^2 \delta - \frac{\rho V_{x\infty}^2 \delta}{3} - \rho V_{x\infty} l \frac{V_{x\infty} \delta}{2l}$$

$$\frac{\mu V_{x\infty} l}{\delta} \approx \frac{\rho V_{z\infty}^2 \delta}{6}$$

Thus

$$\delta^2 \approx \frac{6 \nu l}{V_{x\infty}}$$

Equations (D1) and (D2) can now be simplified by introducing the order of magnitude of each term and then eliminating the small terms. δ is taken as small in comparison with distances in the x direction over which appreciable change occurs, and for the purpose of obtaining orders of magnitude we treat all symbols as dimensionless with respect to l and $V_{x\infty}$. Thus we take

---

Orders of magnitude

V_x and $V_y = 0$ at $y = 0$
δ is taken as small
$Re = V_{x\infty} l / \nu \gg 1$

$$\frac{\partial V_x}{\partial y} \sim 0(\delta^{-1})$$

$$\frac{\partial^2 V_x}{\partial y^2} \sim 0(\delta^{-2})$$

$$V_x, \frac{\partial V_x}{\partial t}, \frac{\partial V_x}{\partial x}, \frac{\partial^2 V_x}{\partial x^2} \sim 0(1)$$

From the continuity equation (D3), since

$$\frac{\partial V_x}{\partial x} \sim 0(1), \quad \therefore \quad \frac{\partial V_y}{\partial y} \sim 0(1)$$

V_y and y increase together so $V_y \sim 0(\delta)$.

Also

$$\frac{\partial V_y}{\partial t}, \frac{\partial V_y}{\partial x}, \frac{\partial^2 V_y}{\partial x^2} \sim 0(\delta)$$

$$\frac{\partial^2 V_y}{\partial y^2} \sim 0(\delta^{-1})$$

It is also assumed that actually or statistically $\partial/\partial t = 0$.

These can now be used to simplify equation (D1).

$$\frac{\partial V_x}{\partial t} + V_x \frac{\partial V_x}{\partial x} + V_y \frac{\partial V_x}{\partial y} = -\frac{1}{\rho}\frac{\partial p}{\partial x} + v\left(\frac{\partial^2 V_x}{\partial x^2} + \frac{\partial^2 V_x}{\partial y^2}\right) \qquad (D1)$$

$$\downarrow \quad \downarrow \quad \downarrow \quad \downarrow \quad \downarrow \quad \downarrow$$

$$0 \qquad 0(1) \quad 0(\delta)x0(\delta^{-1}) \quad \downarrow \quad \frac{1}{Re}(0(1) + 0(\delta^{-2}))$$

$$\downarrow \quad \downarrow \quad \downarrow \quad \text{neglect} \quad \downarrow$$

$$V_x \frac{\partial V_x}{\partial x} + V_y \frac{\partial V_x}{\partial y} = -\frac{1}{\rho}\frac{\partial p}{\partial x} + v\frac{\partial^2 V_x}{\partial y^2}$$

δ is of order $Re^{-\frac{1}{2}}$. The term of $0(1)$ can be neglected in comparison with the term of $0(\delta^{-2})$, since both are within the same brackets. In the boundary layer the viscous and inertia terms are of the same order.

Equation (D2) can now be simplified

$$\frac{\partial V_y}{\partial t} + V_x \frac{\partial V_y}{\partial x} + V_y \frac{\partial V_y}{\partial y} = -\frac{1}{\rho}\frac{\partial p}{\partial y} + v\left(\frac{\partial^2 V_y}{\partial x^2} + \frac{\partial^2 V_y}{\partial y^2}\right) \qquad (D2)$$

$$\downarrow \quad \downarrow \quad \downarrow \quad \downarrow \quad \downarrow \quad \downarrow$$

$$0 \qquad 0(\delta) \qquad 0(\delta) \qquad \downarrow \quad \underbrace{\frac{1}{Re}(0(\delta) + 0(\delta^{-1}))}_{0(\delta)}$$

$$\therefore \quad \frac{1}{\rho}\frac{\partial p}{\partial y} \sim 0(\delta) \text{ or less.}$$

But from the above orders of magnitude $\partial/\partial y = 0(\delta^{-1})$.

∴ pressure change $\sim 0(\delta^2)$.

Thus the outcome of treating the y-component momentum equation in this way is to show that the change in pressure across the boundary layer will be negligible and so the pressure may be treated as of constant value along any normal to the boundary and equal in value to the pressure in the main stream.

The equations for the laminar boundary layer can now be written in their simplified form:

Pressure

pressure, p, is constant across the layer

Continuity

$$\frac{\partial V_x}{\partial x} + \frac{\partial V_y}{\partial y} = 0 \qquad (D3)$$

Momentum

$$V_x \frac{\partial V_x}{\partial x} + V_y \frac{\partial V_x}{\partial y} = -\frac{1}{\rho}\frac{\partial p}{\partial x} + \frac{v\partial^2 V_x}{\partial y^2} \qquad (D4)$$

Boundary conditions

$V_x = V_y = 0$ when $y = 0$.

The last line states that there is a no-slip condition at the boundary and that no flow can pass through the boundary.

Laminar boundary layer on a flat plate – solution

Some further simplification is possible since the pressure gradient along the plate is assumed to be zero.

Pressure

pressure, p, is constant across the layer

$$\frac{\partial p}{\partial x} = 0$$

Continuity

$$\frac{\partial V_x}{\partial x} + \frac{\partial V_y}{\partial y} = 0$$

Momentum

$$V_x \frac{\partial V_x}{\partial x} + V_y \frac{\partial V_x}{\partial y} = \nu \frac{\partial^2 V_x}{\partial y^2} \qquad (D5)$$

Boundary conditions

$$V_x = V_y = 0 \quad \text{when y} = 0$$
$$V_x \to V_{x\infty} \quad \text{as y} \to \infty.$$

A reduction of the partial differential equation (D5) to an ordinary differential equation is possible using the substitution due to Blasius **(57)**

$$\frac{V_x}{V_{x\infty}} = g(y/\Delta)$$

where Δ is an arbitrary length scale (and a function of x only).

If $V_x = \partial \psi / \partial y$ and $V_y = -\partial \psi / \partial x$ then continuity is satisfied and

$$\frac{\psi}{V_{x\infty}\Delta} = f(y/\Delta) \quad \text{when } g = \frac{df}{d\eta} \text{ and } \eta = y/\Delta$$

Substituting in the equation of motion

$$\frac{V_{x\infty}^2}{\Delta}\frac{d\Delta}{dx}\frac{fd^2f}{d\eta^2} + \frac{\nu V_{x\infty}}{\Delta^2}\frac{d^3f}{d\eta^3} = 0$$

If

$$\Delta^2 = \frac{\nu x}{V_{x\infty}}$$

then

$$\frac{d\Delta}{dx} = \frac{1}{2}\left(\frac{\nu}{V_{x\infty}x}\right)^{1/2}$$

and $ff'' + 2f''' = 0$
where $f' = df/d\eta$ etc.

Appendix D: Boundary Layer Equations

and the boundary conditions become

$$f = f' = 0 \quad \text{at } \eta = 0$$
$$f' \to 1 \quad \text{as } \eta \to \infty$$

The solution of this obtained numerically is the Blasius profile (Fig. D3). It has the property that (**57**)

$$\frac{V_x}{V_{x\infty}} = g = f' = 0.99$$

when $\eta = y/\Delta = 4.99$
and if $\delta = y$ when $V_x = 0.99\, V_{x\infty}$ (which is one way of defining the edge of the boundary layer), then

$$\delta = 4.99\, \Delta = 4.99\, (\nu x/V_{x\infty})^{1/2}.$$

Note from the curve in Fig. D3 that for small values of η, $V_x/V_{x\infty}$ is almost proportional to η, and that $V_x/V_{x\infty}$ approaches unity asymptoti-

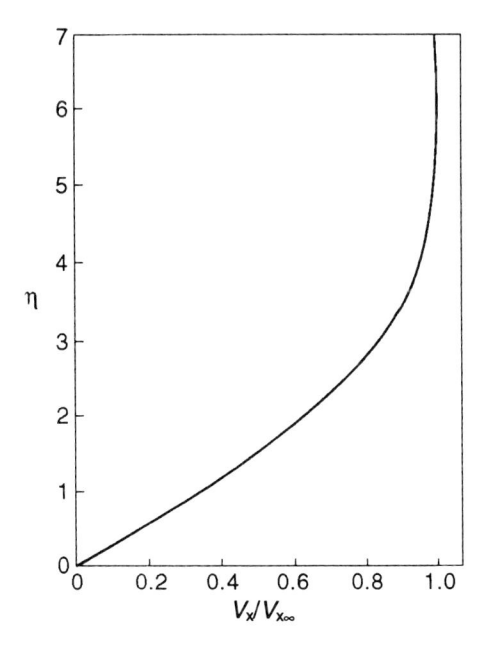

Fig. D3 Blasius profile

cally as $\eta \to \infty$. This could of course, only occur for ideal conditions and an infinite plate.

Oscillating boundary layer

For the boundary layer on an oscillating plate, V_x becomes small, and the viscous term is equated to the time variation of the velocity. Pressure will be essentially constant everywhere and the boundary condition at infinity will be for a stationary fluid. Therefore, (56)

$$\frac{\partial V_x}{\partial t} = v \frac{\partial^2 V_x}{\partial y^2}$$

Boundary condition at the wall:

$V_x = V_{x0} \cos(nt)$ for $y = 0$

This equation is satisfied by

$V_x = V_{x0} e^{-ky} \cos(nt - ky)$

where

$k = \sqrt{(n/2v)}$

Therefore, a wave-like oscillation diffuses outwards with a wave length of $2\pi\sqrt{(2v/n)}$, an amplitude $V_{x0} e^{y\sqrt{(n/2v)}}$ and phase lag $y\sqrt{(n/2v)}$. This is shown in Fig. D4.

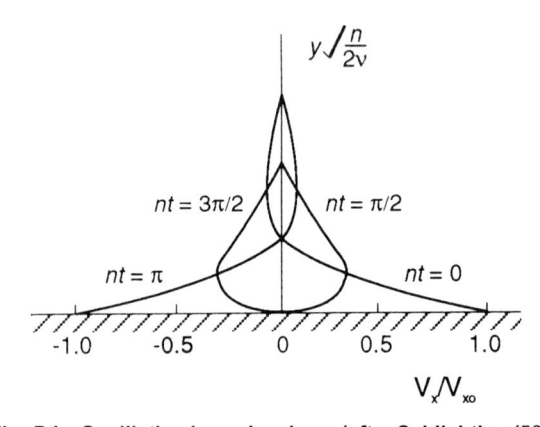

Fig. D4 Oscillating boundary layer (after Schlichting (56))

Flow in narrow passages

Before leaving laminar flows, we consider the flow in a small gap. Such flows are of particular importance in leakage flows and flows past seals or pistons. We assume that the gap is sufficiently long so that the flow has become fully developed and that it is sufficiently wide so that the flow may be treated as two-dimensional. Flow through a parallel passage where one wall of the passage is moving with velocity V_{x0} will obey the equations:

Pressure
   pressure, p, is constant across the passage

Continuity
   We assume that there is no change in the x-direction and, therefore, that there is no flow in the y-direction.

Momentum

$$\nu \frac{\partial^2 V_x}{\partial y^2} = -\frac{1}{\rho}\frac{\partial p}{\partial x} = -\frac{1}{\rho}\frac{p_u - p_d}{L}$$

Boundary conditions

   $V_x = 0$ when $y = 0$.
   $V_x = V_{x0}$ when $y = t$.

The solution of these equations for Fig. D5 is:

$$V_x = \frac{V_{x0} y}{t} + \frac{1}{2\mu}\frac{p_u - p_d}{L} y(t - y)$$

Flow through the gap, per unit width, is

$$q_v = \int_0^t V_x \mathrm{d}y$$

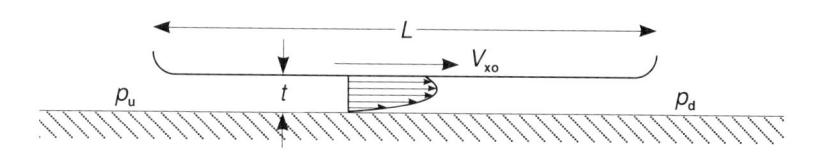

Fig. D5 Flow in a narrow passage

which results in

$$q_v = \frac{V_{x0}t}{2} + \frac{1}{12\mu}\frac{p_u - p_d}{L}t^3$$

Turbulent boundary layers

So far the equations derived have been strictly applicable to laminar flows only. However, it is possible to make them general without much change. To achieve this we first introduce a new description for the variables, velocity and pressure, which recognizes that each variable is the sum of two parts. For the x-component of velocity we use V_x for the mean and V'_x for the fluctuating part which will become zero in laminar flows. For the pressure we use p and p'. If this new set of variables is introduced into the equations, the turbulent boundary layer form can be written as

Momentum

$$(V_x + V'_x)\frac{\partial(V_x + V'_x)}{\partial x} + (V_y + V'_y)\frac{\partial(V_x + V'_x)}{\partial y}$$
$$= -\frac{1}{\rho}\frac{\partial(p + p')}{\partial x} + v\frac{\partial^2(V_x + V'_x)}{\partial y^2}$$

If we take the mean we obtain

$$V_x\frac{\partial V_x}{\partial x} + V_y\frac{\partial V_x}{\partial y} = -\frac{1}{\rho}\frac{\partial p}{\partial x} + v\frac{\partial^2 V_x}{\partial y^2} + \overline{-V'_x\frac{\partial V'_x}{\partial x} - V'_y\frac{\partial V'_x}{\partial y}} \qquad \text{(D6)}$$

Much of this equation is the same as for the laminar conditions. Were it possible to rearrange the final term, we might have a useful and easily soluble equation. We can move in this direction by using the continuity equation.

Continuity

$$\frac{\partial(V_x + V'_x)}{\partial x} + \frac{\partial(V_y + V'_y)}{\partial y} = 0$$

but

$$\frac{\partial V_x}{\partial x} + \frac{\partial V_y}{\partial y} = 0$$

Appendix D: Boundary Layer Equations

Flow in narrow passages
Before leaving laminar flows, we consider the flow in a small gap. Such flows are of particular importance in leakage flows and flows past seals or pistons. We assume that the gap is sufficiently long so that the flow has become fully developed and that it is sufficiently wide so that the flow may be treated as two-dimensional. Flow through a parallel passage where one wall of the passage is moving with velocity V_{x0} will obey the equations:

Pressure
  pressure, p, is constant across the passage

Continuity
  We assume that there is no change in the x-direction and, therefore, that there is no flow in the y-direction.

Momentum
$$\nu \frac{\partial^2 V_x}{\partial y^2} = -\frac{1}{\rho}\frac{\partial p}{\partial x} = -\frac{1}{\rho}\frac{p_u - p_d}{L}$$

Boundary conditions
  $V_x = 0$ when $y = 0$.
  $V_x = V_{x0}$ when $y = t$.

The solution of these equations for Fig. D5 is:
$$V_x = \frac{V_{x0} y}{t} + \frac{1}{2\mu}\frac{p_u - p_d}{L} y(t - y)$$

Flow through the gap, per unit width, is
$$q_v = \int_0^t V_x \, dy$$

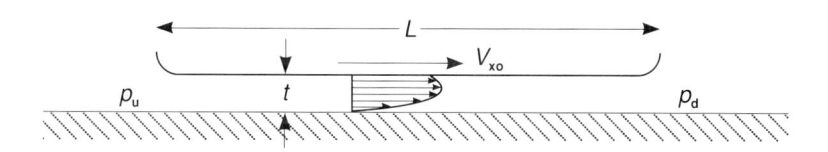

Fig. D5 Flow in a narrow passage

which results in

$$q_v = \frac{V_{x0}t}{2} + \frac{1}{12\mu}\frac{p_u - p_d}{L}t^3$$

Turbulent boundary layers
So far the equations derived have been strictly applicable to laminar flows only. However, it is possible to make them general without much change. To achieve this we first introduce a new description for the variables, velocity and pressure, which recognizes that each variable is the sum of two parts. For the x-component of velocity we use V_x for the mean and V'_x for the fluctuating part which will become zero in laminar flows. For the pressure we use p and p'. If this new set of variables is introduced into the equations, the turbulent boundary layer form can be written as

Momentum

$$(V_x + V'_x)\frac{\partial(V_x + V'_x)}{\partial x} + (V_y + V'_y)\frac{\partial(V_x + V'_x)}{\partial y}$$
$$= -\frac{1}{\rho}\frac{\partial(p + p\prime)}{\partial x} + v\frac{\partial^2(V_x + V'_x)}{\partial y^2}$$

If we take the mean we obtain

$$V_x\frac{\partial V_x}{\partial x} + V_y\frac{\partial V_x}{\partial y} = -\frac{1}{\rho}\frac{\partial p}{\partial x} + v\frac{\partial^2 V_x}{\partial y^2} + \overline{-V'_x\frac{\partial V'_x}{\partial x} - V'_y\frac{\partial V'_x}{\partial y}} \qquad \text{(D6)}$$

Much of this equation is the same as for the laminar conditions. Were it possible to rearrange the final term, we might have a useful and easily soluble equation. We can move in this direction by using the continuity equation.

Continuity

$$\frac{\partial(V_x + V'_x)}{\partial x} + \frac{\partial(V_y + V'_y)}{\partial y} = 0$$

but

$$\frac{\partial V_x}{\partial x} + \frac{\partial V_y}{\partial y} = 0$$

so

$$\frac{\partial V'_x}{\partial x} + \frac{\partial V'_y}{\partial y} = 0$$

Applying this to the final terms in equation (D6)

$$\overline{-V'_x \frac{\partial V'_x}{\partial x} - V'_y \frac{\partial V'_x}{\partial y}}$$

this becomes

$$\overline{\frac{-\partial V'^2_x}{\partial x} + V'_x \frac{\partial V'_x}{\partial x} - V'_y \frac{\partial V'_x}{\partial y}}$$

$$= \overline{\frac{-\partial V'^2_x}{\partial x} - V'_x \frac{\partial V'_y}{\partial y} - V'_y \frac{\partial V'_x}{\partial y}}$$

$$= \overline{\frac{-\partial V'^2_x}{\partial x} - \frac{\partial (V'_x V'_y)}{\partial y}}$$

The first term can be neglected in comparison with the second. The equation for the turbulent boundary layer is then

$$V_x \frac{\partial V_x}{\partial x} + V_y \frac{\partial V_x}{\partial y} = \frac{1}{\rho}\frac{\partial p}{\partial x} + \frac{\partial}{\partial y}\left(v \frac{\partial V_x}{\partial y} - \overline{V'_x V'_y} \right) \tag{D7}$$

If we could write

$$-\overline{V'_x V'_y} = \frac{\tau}{\rho} = \frac{\epsilon}{\rho}\frac{\partial V_x}{\partial y}$$

then the last term in the boundary layer equation becomes

$$\frac{\epsilon + \mu}{\rho}\frac{\partial^2 V_x}{\partial y^2}$$

The new 'eddy' viscosity is found to be larger than the kinematic viscosity term for turbulent regions. Notable fluid mechanicists like Prandtl and G I Taylor have suggested a simple mixing length theory to obtain the value of the eddy viscosity. Figure D6 shows the main features of this. It is essentially a dimensional argument to identify the parameters affecting the value of τ.

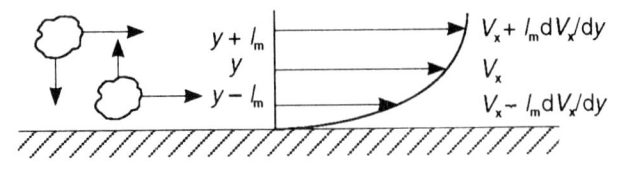

Fig. D6 Diagram to explain mixing length theory

Fluid particles are assumed to move perpendicular to the boundary causing high velocity, $V_x + l_m dV_x/dy$, and low velocity fluid, $V_x - l_m dV_x/dy$, to mix. It is assumed that the expression $l_m dV_x/dy$ is equivalent to the fluctuating turbulent component, V'_x. A strong negative correlation between V'_x and V'_y is assumed since positive V'_x will be associated with negative V'_y. If it is assumed that V'_x and V'_y are of the same order a value is obtained for τ with unknown mixing length, l_m.

$$\tau = -\rho V'_x V'_y = \rho l_m^2 \left|\frac{dV_x}{dy}\right|\left(\frac{dV_x}{dy}\right)$$

Launder and Spalding (**58**) give the following values for mixing lengths:

Plane mixing layer:
$l_m/\delta = 0.07$ (δ = layer width)
Plane jet in stagnant surroundings:
$l_m/\delta = 0.09$ (δ = width of half jet)
Fan jet in stagnant surroundings:
$l_m/\delta = 0.125$ (δ = width of half jet)
Round jet in stagnant surroundings:
$l_m/\delta = 0.075$ (δ = width of half jet)
Turbulent pipe flow:
$l_m/R = 0.14 - 0.08(1 - y/R)^2 - 0.06(1 - y/R)^4$

Displacement thickness
A very useful approach to the solution of flow fields where the effect of boundary layers needs to be taken into account is to solve the two regions separately. Thus the thickness of the boundary layer is calculated and the flow field is then given an artificial boundary, displaced by the thickness of the boundary layer. This thickness, δ^*, is calculated as though there were no flow in the boundary layer and a slip condition at the artificial

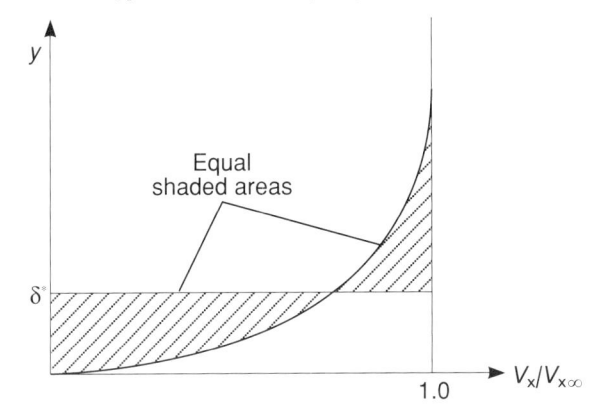

Fig. D7 Diagram to show displacement thickness

boundary. Having calculated the flow field, it is then possible to obtain a revised boundary layer thickness based on the actual pressure distribution in the free stream. Figure D7 shows the integration required to obtain the displacement thickness and the equations for this are:

Displacement thickness

$$\delta^* = \int_0^\infty \left(1 - \frac{V_x}{V_{x\infty}}\right) dy$$

Momentum thickness

$$\theta = \int_0^\infty \frac{V_x}{V_{x\infty}} \left(1 - \frac{V_x}{V_{x\infty}}\right) dy$$

Energy thickness

$$\delta_E = \int_0^\infty \frac{V_x}{V_{x\infty}} \left(1 - \left(\frac{V_x}{V_{x\infty}}\right)^2\right) dy$$

The values for δ^* and θ can be obtained by using the approximate expression for the turbulent boundary layer shape on a flat plate

$$\frac{V_x}{V_{x\infty}} = \left(\frac{y}{\delta}\right)^{1/n}$$

where n is often taken as 7.

General expression		1/n		1/7
$\delta^* = \int_0^\infty \left(1 - \dfrac{V_x}{V_{x\infty}}\right) dy$	=	$\dfrac{\delta}{n+1}$	=	$\dfrac{\delta}{8}$
$\theta = \int_0^\infty \dfrac{V_x}{V_{x\infty}}\left(1 - \dfrac{V_x}{V_{x\infty}}\right) dy$	=	$\dfrac{n\delta}{(n+1)(n+2)}$	=	$\dfrac{7\delta}{72}$

Two, of many, criteria which aim to predict boundary layer separation, all using data correlations, are:

(a) a zero pressure gradient turbulent boundary layer on a flat plate does not separate;
(b) the shape factor of the boundary layer is $H = \delta^*/\theta$ and Schlichting (**56**) shows that separation occurs when $H > 2 \to 2.5$. In the boundary layer shape above, the value of $H = 1.3$. Using the expression above, if $n = 2$ then $H = 2$, and if $n = 1$ (Fig. 5.3) then $H = 3$ (cf Fig. 5.9).

APPENDIX E

Derivation of the Compressible Flow Equations

In this appendix the derivation of the compressible flow equations is provided (**33**) from basic equations which have been referred to in the main text:

1. Continuity – as in equation (3.13)

$$q_m = \rho_1 A_1 V_1 = \rho_2 A_2 V_2 \tag{E1}$$

2. Energy – as in equation (3.19)

$$Q - W = q_m(h_2 + \frac{V_2^2}{2} + gz_2) - q_m(h_1 + \frac{V_1^2}{2} + gz_1) \tag{E2}$$

3. Perfect gas equation

$$pv = RT \tag{E3}$$

and equation (3.18)

$$h_2 - h_1 = c_p(T_2 - T_1) \tag{E4}$$

4. Isentropic relationship, equation (3.22)

$$Tds = dh - vdp \tag{E5}$$

and the integrated form of this for a perfect gas, equation (3.24)

$$s_2 - s_1 = c_p ln(T_2/T_1) - Rln(p_2/p_1) \tag{E6}$$

$R = c_p - c_v$, and the symbol, γ, are also needed for the ratio of the specific heats where $\gamma = c_p/c_v$.

Expression for the speed of sound

It is useful first of all to obtain the expression for the speed of sound by applying the momentum equation across a sound wave as in Fig. E1. Applying the momentum equation across a stationary sound wave in a duct

$$A(p - (p + dp)) = q_m((c - dv) - c)$$

An Introductory Guide to Industrial Flow

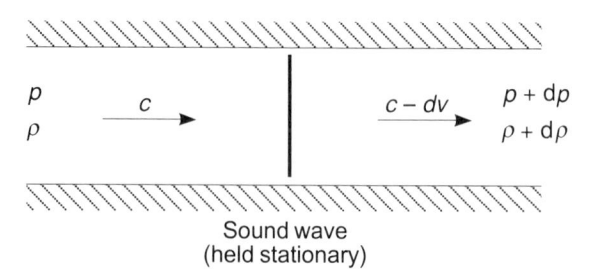

Fig. E1 Sound wave momentum balance

which after simplifying becomes

$$Adp = q_m dv \tag{E7}$$

Using equation (E1) this may be rewritten as

$$dp = \rho c\, dv \tag{E8}$$

Applying continuity across the sound wave we obtain

$$\rho A c = (\rho + d\rho) A (c - dv)$$

which may be rewritten

$$\rho c = \rho c + c\, d\rho - \rho\, dv$$

which gives

$$c\, d\rho = \rho\, dv \tag{E9}$$

Combining equations (E8) and (E9) we obtain

$$c^2 = \frac{dp}{d\rho}$$

and if we assume that the process of a sound wave is isentropic we may write

$$c^2 = \left(\frac{\partial p}{\partial \rho}\right)_s \tag{E10}$$

Appendix E: Derivation of the Compressible Flow Equations

From equations (E3) and (E6) it can be shown that for a perfect gas

$$\frac{dp}{p} = \gamma \frac{d\rho}{\rho} \tag{E11}$$

and hence for a perfect gas the speed of sound is given by

$$c^2 = \frac{\gamma p}{\rho} = \gamma RT \tag{E12}$$

and the Mach number is given by

$$M = \frac{V}{\sqrt{(\gamma RT)}} \tag{E13}$$

Compressible flow equations

Equation (6.1) can now be obtained since from equation (E2) we have (for $Q = W = 0$ and neglecting terms in z)

$$h_0 = h + \frac{V^2}{2} \tag{E14}$$

or for a perfect gas

$$T_0 = T + \frac{V^2}{2c_p} \tag{E15}$$

which can be rewritten

$$\frac{T_0}{T} = 1 + \frac{\gamma - 1}{2} M^2 \tag{6.1}$$

Equation (6.2) can be obtained from this one using equation (E6) (with $s_2 = s_1$), and equation (6.4) may be obtained from the continuity equation (E1)

$$\frac{q_m}{A} = \rho V = \frac{p}{RT} V = \frac{p}{p_0} \frac{V}{\sqrt{(\gamma RT)}} \sqrt{\frac{\gamma}{R}} \sqrt{\frac{T_0}{T}} \sqrt{\frac{p_0}{T_0}}$$

which after manipulation becomes

$$q_m \frac{\sqrt{(c_p T_0)}}{A p_0} = \frac{\gamma}{\sqrt{(\gamma - 1)}} M \left(1 + \frac{\gamma - 1}{2} M^2\right)^{-\frac{1}{2}(\gamma+1)/(\gamma-1)} \tag{6.4}$$

Similar manipulation yields equations (6.3) and (6.5).

Shock wave equations
Using Fig. E2 we obtain:

Continuity from equation (E1)

$$\rho V = \rho_s V_s \tag{E16}$$

Energy

$$c_p T + \frac{V^2}{2} = c_p T_s + \frac{V_s^2}{2} \tag{E17}$$

Momentum

$$p + \rho V^2 = p_s + \rho V_s^2 \tag{E18}$$

Equation (E16) can be rewritten since $V = M\sqrt{(\gamma p/\rho)}$ and $p/\rho = RT$ as

$$\frac{M^2 p^2}{T} = \frac{M_s^2 p_s^2}{T_s} \tag{E19}$$

Equation (E17) can be rewritten since $V^2 = M^2 \gamma RT$ as

$$T\{1 + \tfrac{1}{2} M^2(\gamma - 1)\} = T_s\{1 + \tfrac{1}{2} M_s^2(\gamma - 1)\} \tag{E20}$$

and equation (E18) can be rewritten since $V^2 = M^2 \gamma p/\rho$ as

$$p(1 + \gamma M^2) = p_s(1 + \gamma M_s^2) \tag{E21}$$

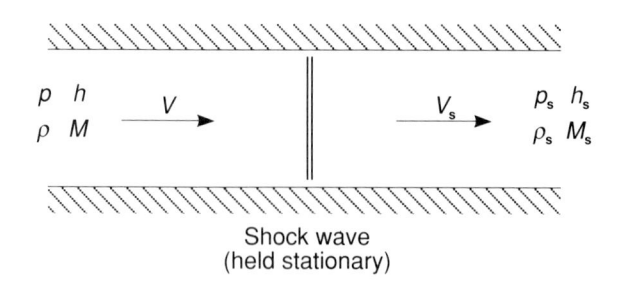

Fig. E2 Shock wave momentum balance

Appendix E: Derivation of the Compressible Flow Equations 215

So eliminating p, T, p_s, and T_s we obtain

$$\frac{M^2\{1 + \frac{1}{2}M^2(\gamma - 1)\}}{(1 + \gamma M^2)^2} = \frac{M_s^2\{1 + \frac{1}{2}M_s^2(\gamma - 1)\}}{(1 + \gamma M_s^2)^2} \tag{E20}$$

And using the standard solution for a quadratic equation to obtain the roots of M_s^2 equation (6.6) is obtained, since $M_s^2 = M^2$ is a trivial solution

$$M_s^2 = \frac{1 + \frac{\gamma - 1}{2} M^2}{\gamma M^2 - \frac{\gamma - 1}{2}}$$

REFERENCES

(**1**) R. C. Baker, *An introductory guide to flow measurement*, 1989, (Mechanical Engineering Publications, London).
(**2**) H. Lamb, *The dynamical theory of sound*, 1960, (Dover Publications Inc.).
(**3**) R. C. Baker, 'An engineering program for centrifugal compressors: flow visualization tests of impellers' Reports 1143–23 and 1143–31, 1970 and 1971, (Northern Research and Engineering Corporation, Mass, USA).
(**4**) K. Ahmad, R. C. Baker and A. Goulas, 'Computation and experimental results of wear in a slurry pump impeller', *Proc. Instn. Mech. Engrs*, 1986, **200** (C6).
(**5**) R. C. Baker, 'Electromagnetic Wall Velometers', PhD Thesis, 1967, University of Cambridge.
(**6**) R. C. Baker, 'The design of an electromagnetic ship's log which causes minimal obstruction outside the hull', Unpublished report, 1968.
(**7**) S. Goldstein, *Modern developments in fluid dynamics*, 1965, (Dover Publications Inc).
(**8**) A. Goulas and R. C. Baker, 'Flows at the exit of a centrifugal compressor impeller', *Proc. Instn. Mech. Engrs*, 1979, **193**, (33), 341–347
(**9**) J. A. Damia Torres, A time marching, finite area calculation of general quasi-one-dimensional flows, PhD Thesis, 1979, Imperial College, London.
(**10**) C. P. Lenn, J. Hemp, R. C. Baker, E. R. Hayes and A. D. Harper, 'A study of flow downstream of a T-junction and the implications for its effect (in crude oil flows) on water droplet size and distribution', *Proc. Instn. Mech. Engrs*, 1993.
(**11**) B. E. Noltingk, (Editor), *Instrumentation reference book*, 1988, (Butterworths, London).
(**12**) C. Hagart-Alexander, *Instrumentation reference book*, 1988, (Edited by B. E. Noltingk), (Butterworths, London).
(**13**) D. S. Miller, *Internal flow systems*, 1990, Second edition, (Gulf Publishing Company, London).

(14) E. H. Higham, *Instrumentation reference book*, 1988, (Edited by B. E. Noltingk), (Butterworths, London).
(15) G. W. C. Kaye and T. H. Laby, *Tables of physical and chemical constants*, 1956, (Longmans).
(16) D. A. Kelly, *Instrumentation reference book*, 1988, (Edited by B. E. Noltingk), (Butterworths, London).
(17) K. Walters and W. M. Jones, *Instrumentation reference book*, 1988, (Edited by B. E. Noltingk), (Butterworths, London).
(18) R. W. Miller, *Flow measurement engineering handbook*, 1989, Second edition, (McGraw-Hill, New York).
(19) W. J. Duncan, A. S. Thom and A. D. Young, *Mechanics of fluids*, 1960, Edward Arnold.
(20) R. C. Baker, 'Computational method to assess concentration of water in crude oil downstream of a mixing section', *Proc. Inst. Mech. Engrs*, 1988, **202** (A2), 117–127.
(21) J. K. Reichert and R. S. Azad, 'Features of a developing turbulent boundary layer measured in a bounded flow', *Can. J. Phys.*, 1979, **57**, 477–485.
(22) J. H. Keenan, *Thermodynamics*, 1970, (The MIT Press).
(23) R. W. Hayward, *Equilibrium thermodynamics for engineers and scientists*, 1980, (John Wiley, Chichester).
(24) A. J. Ward-Smith, *Internal fluid flow*, 1980, (Oxford University Press).
(25) I. P. Bates, 'Field use of K-lab flow conditioner', *Proceedings of the North Sea Flow Measurement Workshop*, 1991, (Norwegian Society of Chartered Engineers).
(26) U. Karrick, W. M. Jungowski and K. Botros, 'Effects of flow characteristics downstream of elbow/flow conditioner on orifice meter accuracy', *Proceedings of the North Sea Flow Measurement Workshop*, 1991, (Norwegian Society of Chartered Engineers).
(27) L. Prandtl, *Verhandlungen des dritten Internationalen Mathematiker Kongresses*, 1904, 484–491.
(28) J. M. Kay, *An introduction to fluid mechanics and heat transfer*, 1968, Second edition, (Cambridge University Press).
(29) Meteorological Office, *Meteorology for mariners*, 1978, Third edition, (HMSO).

(30) L. Prandtl and O. G. Tietjens, *Fundamentals of hydro- and aeromechanics and applied hydro- and aeromechanics*, 1957, (Dover Publications Inc.).
(31) M. Van Dyke, *An album of fluid motion*, 1982, (The Parabolic Press, California, USA).
(32) J. D. Summers-Smith, *An introductory guide to industrial tribology*, 1994, (Mechanical Engineering Publications, London).
(33) A. H. Shapiro, *The dynamics and thermodynamics of compressible fluid flow*, 1953, (The Ronald Press Company, New York).
(34) R. C. Asher, 'Ultrasonic transducers for chemical and process plant', *Physics Technology*, 1983, **14**.
(35) R. W. Herschy, *Streamflow measurement*, 1995, Second edition, (E. & F. N. Spon).
(36) R. C. Baker and E. R. Hayes, 'Multiphase measurement problems and techniques for oil production systems', *Petroleum Review*, 1985.
(37) R. C. Baker, 'Response of bulk flowmeters to multiphase flows', *Proc. Instn. Mech. Engrs*, 1991, **205** 217–229.
(38) D. Butterworth and G. F. Hewitt, *Two-phase flow and heat transfer*, 1977, Harwell Series, (Oxford University Press).
(39) G. Hetsroni, *Handbook of multiphase systems*, 1981, (McGraw-Hill).
(40) N. H. Thomas, T. R. Auton, K. Sene and J. C. R. Hunt, 'Entrapment and transport of bubbles by transient large eddies in multiphase turbulent shear flow', *International conference on physical modelling of multi-phase flow*, 1983, Paper E1, (BHRA Fluid Engineering, Cranfield), 169–184.
(41) R. C. Baker and J. E. Deacon, 'Tests on turbine, vortex and electromagnetic flowmeters in 2 phase air/water upward flow', *International conference on physical modelling of multi-phase flow*, 1983, Paper H1, (BHRA Fluid Engineering, Cranfield), 337–352.
(42) J.-P. Hulin and A. J. M. Foussat, 'Vortex flowmeter behaviour in liquid–liquid two-phase flow', *International conference on physical modelling of multi-phase flow*, 1983, Paper H3, (BHRA Fluid Engineering, Cranfield), 377–390.
(43) S. A. Morsi and A. J. Alexander, 'An investigation of particle trajectories in two-phase flow systems', *J. Fluid Mech*, 1972, **55**, 193–208.

(44) R. Clift, J. R. Grace and M. E. Weber, *Bubbles, drops and particles*, 1978, (Academic Press, New York).
(45) R. A. Budenholzer, 'Steam', *Encyclopaedia Britannica*, 1971, **21**, 175.
(46) M. W. Zemansky, *Heat and thermodynamics*, 1957, Fourth edition, (McGraw-Hill).
(47) R. W. Hayward, *Thermodynamic tables in SI (metric) units*, 1968, (Cambridge University Press).
(48) D. C. Hickson and F. R. Taylor, Enthalpy–entropy diagram for steam, 1980, (Basil Blackwell, Oxford).
(49) A. Goulas and R. C. Baker, 'Through flow analysis of viscous and turbulent flows', ARC 37 017, 1976.
(50) N. D. Vaughan, D. N. Johnston and K. A. Edge, 'Numerical simulation of fluid flow in poppet valves', *J. Mech. Engng Sci*, 1992, **206**, 119–127.
(51) P. Koutmos and J. J. McGuirk, 'Numerical calculations of the flow in annular combustor dump diffuser geometries', *J. Mech. Engng Sci*, 1989, **203**, 319–331.
(52) A. K. El Wahed, M. W. Johnson and J. L. Sproston, 'Numerical study of vortex shedding from different shaped bluff bodies', *Flow Measurement and Instrumentation*, 1993, **4**, 233–240.
(53) U. Buckle, F. Durst, B. Howe and A. Melling, 'Investigation of a floating element flowmeter', *Flow Measurement and Instrumentation*, 1992, **3**, 215–225.
(54) Y. Xu, 'Calculation of the flow around turbine flowmeter blades', *Flow Measurement and Instrumentation*, 1992, **3**, 25–35.
(55) J. H. Horlock, 'Approximate equations for the properties of superheated steam', *Proc. Instn Mech. Engrs*, 1959, **173**, 779–794.
(56) H. Schlichting, *Boundary layer theory*, 1955, (McGraw Hill).
(57) D. J. Tritton, *Physical fluid dynamics*, 1988, Second edition, (Oxford University Press).
(58) B. E. Launder and D. B. Spalding, *Mathematical models of turbulence*, 1972, (Academic Press).

BIBLIOGRAPHY

The books and papers below are a selection which I have found particularly useful or which give a recent overview of an area of the subject. I have included some, such as those by Bagnold and Shercliff, which take the subject well beyond the scope of this book, but which are valuable texts for those who wish to probe these particular aspects of the subject.

Flow visualization
Pictures of flows are worth many words and equations and both these books provide excellent examples.

Japan Society of Mechanical Engineers, *Visualized flow–fluid motion in basic and engineering situations revealed by flow visualization*, 1988, (Pergamon Press).

M. **Van Dyke**, *An album of fluid motion*, 1982, (The Parabolic Press, California, USA).

Fluid parameters
The first of these books is a good example of a handbook produced by a manufacturer, while the second is a very extensive book covering many aspects of instrumentation.

H. **Julien**, *WIKA Handbook of pressure measurement with resilient elements*, 1981, (Alexander Wiegand GmbH).

B. E. **Noltingk** (Editor), *Instrumentation reference book*, 1988, (Butterworths, London).

Flow measurement
Miller's book is a thorough treatment of the subject with an emphasis on differential pressure methods. Herschy is one of the international experts on open-channel flowmeasurement. My own book and that produced by Flowtec provide brief presentations, the latter more from the industrial viewpoint. I include my paper since it provides a recent review of an important category of flow measurement.

R. C. **Baker**, *An introductory guide to flow measurement*, 1988, (Mechanical Engineering Publications, London).

R. C. Baker, 'Response of bulk flowmeters to multiphase flows', *Proc. Instn. Mech. Engrs*, 1991, **205**, 217–229.

R. W. Herschy, *Streamflow measurement*, 1995, Second edition (E. & F. N. Spon).

R. W. Miller, *Flow measurement engineering handbook*, 1989, Second edition (McGraw-Hill, New York).

Endress and Hauser, *Flow handbook*, 1989, (Flowtec AG), English edition.

Fluid mechanics

Acheson and Tritton are two excellent and modern treatments of the subject. I have attempted to emulate Acheson's ability to link the subject with the immediately observable. Kay is a treatment which is very accessible. The others are notable texts.

D. J. Acheson, *Elementary fluid dynamics*, 1990, (Oxford University Press).

W. J. Duncan, A. S. Thom and A. D. Young, *Mechanics of fluids*, 1960, (Howard Arnold).

S. Goldstein, *Modern developments in fluid dynamics*, 1985, (Dover Publications Inc.).

J. M. Kay, *An introduction to fluid mechanics and heat transfer*, 1968, (Cambridge University Press).

L. Prandtl and O. G. Tietjens, *Fundamentals of hydro- and aeromechanics and applied hydro- and aeromechanics*, 1957, (Dover Publications Inc.).

D. J. Tritton, *Physical fluid dynamics*, 1988, Second edition, (Oxford University Press).

Thermodynamics

This selection is based more on my own past use and does not pretend to be an up-to-date review of the available texts.

R. W. Hayward, *Thermodynamic tables in SI (metric) units*, 1968, (Cambridge University Press).

R. W. Hayward, *Equilibrium thermodynamics for engineers and scientists*, 1980, (John Wiley, Chichester).

J. H. Keenan, *Thermodynamics*, 1970, (The MIT Press).

Zemansky, *Heat and thermodynamics*, 1957, Fourth edition, (McGraw-Hill).

BIBLIOGRAPHY

The books and papers below are a selection which I have found particularly useful or which give a recent overview of an area of the subject. I have included some, such as those by Bagnold and Shercliff, which take the subject well beyond the scope of this book, but which are valuable texts for those who wish to probe these particular aspects of the subject.

Flow visualization
Pictures of flows are worth many words and equations and both these books provide excellent examples.

Japan Society of Mechanical Engineers, *Visualized flow–fluid motion in basic and engineering situations revealed by flow visualization*, 1988, (Pergamon Press).

M. Van Dyke, *An album of fluid motion*, 1982, (The Parabolic Press, California, USA).

Fluid parameters
The first of these books is a good example of a handbook produced by a manufacturer, while the second is a very extensive book covering many aspects of instrumentation.

H. Julien, *WIKA Handbook of pressure measurement with resilient elements*, 1981, (Alexander Wiegand GmbH).

B. E. Noltingk (Editor), *Instrumentation reference book*, 1988, (Butterworths, London).

Flow measurement
Miller's book is a thorough treatment of the subject with an emphasis on differential pressure methods. Herschy is one of the international experts on open-channel flowmeasurement. My own book and that produced by Flowtec provide brief presentations, the latter more from the industrial viewpoint. I include my paper since it provides a recent review of an important category of flow measurement.

R. C. Baker, *An introductory guide to flow measurement*, 1988, (Mechanical Engineering Publications, London).

R. C. Baker, 'Response of bulk flowmeters to multiphase flows', *Proc. Instn. Mech. Engrs*, 1991, **205**, 217–229.

R. W. Herschy, *Streamflow measurement*, 1995, Second edition (E. & F. N. Spon).

R. W. Miller, *Flow measurement engineering handbook*, 1989, Second edition (McGraw-Hill, New York).

Endress and Hauser, *Flow handbook*, 1989, (Flowtec AG), English edition.

Fluid mechanics

Acheson and Tritton are two excellent and modern treatments of the subject. I have attempted to emulate Acheson's ability to link the subject with the immediately observable. Kay is a treatment which is very accessible. The others are notable texts.

D. J. Acheson, *Elementary fluid dynamics*, 1990, (Oxford University Press).

W. J. Duncan, A. S. Thom and A. D. Young, *Mechanics of fluids*, 1960, (Howard Arnold).

S. Goldstein, *Modern developments in fluid dynamics*, 1985, (Dover Publications Inc.).

J. M. Kay, *An introduction to fluid mechanics and heat transfer*, 1968, (Cambridge University Press).

L. Prandtl and O. G. Tietjens, *Fundamentals of hydro- and aeromechanics and applied hydro- and aeromechanics*, 1957, (Dover Publications Inc.).

D. J. Tritton, *Physical fluid dynamics*, 1988, Second edition, (Oxford University Press).

Thermodynamics

This selection is based more on my own past use and does not pretend to be an up-to-date review of the available texts.

R. W. Hayward, *Thermodynamic tables in SI (metric) units*, 1968, (Cambridge University Press).

R. W. Hayward, *Equilibrium thermodynamics for engineers and scientists*, 1980, (John Wiley, Chichester).

J. H. Keenan, *Thermodynamics*, 1970, (The MIT Press).

Zemansky, *Heat and thermodynamics*, 1957, Fourth edition, (McGraw-Hill).

Internal flows and flow losses
These are both excellent and valuable books. The first is aimed at the engineer who wishes to calculate losses in a system. The second is a more academic treatment.

D. S. Miller, *Internal Flow Systems*, 1990, Second edition, (Gulf Publishing Company, London).

A. J. Ward-Smith, *Internal fluid flow*, 1980, (Oxford University Press).

Boundary layers
This has for so long been the standard text that it is natural to include it here.

H. Schlichting, *Boundary layer theory*, 1955, (McGraw-Hill).

Compressible flow
I have always found Shapiro clear and thorough, although there are many other treatments of compressible flow. Miller provides an engineers guide to the prediction of compressible flows.

A. H. Shapiro, *The dynamics and thermodynamics of compressible fluid flows*, 1953, (The Ronald Press Company, New York).

D. S. Miller, *Internal flow systems*, 1990, Second edition, (Gulf Publishing Company, London).

Oscillations and Waves
Lamb and Rayleigh are classics and worth perusing, while Morse and Ingard is a thorough modern treatment.

H. Lamb, *The dynamical theory of sound*, 1960, (Dover Publications Inc.).

P. M. Morse and K. U. Ingard, *Theoretical acoustics*, 1968, (McGraw-Hill).

J. W. S. Rayleigh, *The theory of sound*, 1945, (Dover Publications Inc.).

Multiphase flow
The first of these comes from Harwell where much work has been done on multiphase flows. Bagnold is a classic, and the others cover other aspects of the subject. Again this represents the author's preferences and does no justice to the huge literature on the subject.

D. Butterworth and G. F. Hewitt, *Two-phase flow and heat transfer*, 1977, Harwell Series (Oxford University Press).

R. Clift, J. R. Grace and M. E. Weber, *Bubbles, drops and particles*, 1978, (Academic Press, New York).

G. Hetsroni, *Handbook of multiphase systems*, 1981, (McGraw-Hill).

R. A. Bagnold, *The physics of blown sand and desert dunes*, 1941, (Chapman & Hall).

Steam

R. A. Budenholzer, 'Steam', *Encyclopaedia Britannica*, 1971, **21**, 175.

R. W. Hayward, *Thermodynamic tables in SI (metric) units*, 1968, (Cambridge University Press).

D. C. Hickson and F. R. Taylor, *Enthalpy–entropy diagram for steam*, 1980, (Basil Blackwell, Oxford).

General information

J. C. Anderson, D. M. Hum, B. G. Neal and J. H. Whitelaw, *Data and formulae for engineering students*, 1967, (Pergamon Press).

G. W. C. Kaye and T. H. Laby, *Tables of physical and chemical constants*, 1966, (Longmans).

Related areas of fluid engineering

These are largely outside the scope of this book, but provide the reader with follow-up guides if any of these topics are of interest.

E. Smith and B. E. Vivian, *An introductory guide to valve selection*, 1995, (Mechanical Engineering Publications, London).

J. D. Summers-Smith, *An introductory guide to industrial tribology*, 1994, (Mechanical Engineering Publications, London).

R. K. Turton, *An introductory guide to pumps and pumping systems*, 1993, (Mechanical Engineering Publications, London).

R. K. Turton, *Principles of turbomachinery*, 1984, (E & F N Spon).

S. L. Dixon, *Fluid mechanics, thermodynamics of turbomachinery*, 1966, (Pergamon Press).

J. A. Shercliff, *A textbook of magnetohydrodynamics*, 1965, (Pergamon Press).

INDEX

Acoustic impedance 128
Adiabatic 69
Archimedes 53

Beattie–Bridgman equation 157
Bernoulli's equation 29, 53, 71
Blasius profile 202–204
Boundary conditions 201, 202
Boundary layer 83, 85, 87, 197
  edge of 203
  energy thickness 209
  fully developed profile 58
  fully turbulent region 88
  growth of laminar boundary layer 85
  laminar boundary layer 85
  laminar sub-layer 88
  momentum thickness 209
  on a flat plate 84
  on an oscillating plate 204
  outer region 88
  parabolic profile 58, 60
  separation 88, 89, 90, 210
  separation criteria 210
  shape 84
  shape factor 210
  stall 88
  thickness 85, 86
  turbulent boundary layer 86–88, 206, 207
  transition region 85
  transitory stall 88
Bourdon tube 22, 23, 33, 34
  helical 34
  spiral 34
Breakup of droplets 167
Bubbles 140, 142
  coalescence of 141
Bulk flowmeters/flow measurement 13, 58

averaging pitot tube 142
Coriolis 13, 16, 18
critical nozzle venturi flowmeter 93, 101
Dall tube 142
electromagnetic 13, 16, 17
fluidic 89, 121, 125
orifice 13, 15, 17, 142
positive displacement 13, 15, 17
thermal 13, 16, 18
turbine 13, 15, 17, 141
ultrasonic – correlation 128
ultrasonic – doppler 13, 16, 18
ultrasonic – transit time 13, 16, 18, 127, 141
variable area 13, 15, 17
venturi 13, 15, 17
vortex 13, 16, 17
Bulk modulus 124

Cavitation 139
Centre of buoyancy 55
CFD see Computational fluid dynamics
Compressibility 38
Compressible flow 93
  constant area duct with friction – Fanno line 93, 100, 113, 114, 115
  constant area duct with heat transfer – Rayleigh line 93, 113, 118, 119
  convergent duct/nozzle 88, 93–95, 101, 102, 159
  convergent–divergent duct/nozzle 93, 95, 97, 101, 103, 105, 110, 160, 166
  equations 98–100, 211, 213
Computational fluid dynamics 165
Concentration measurement 19, 46

Coriolis flowmeter   18
Couette flow   91
Critical depth   132
Critical isotherm   151
Critical nozzle venturi flowmeter   93, 101
Critical point   151, 152

Dall tube   142
Density   19, 37–39, 149
   relative   37
Density measurement   37
   displacer   39
   gamma ray   48
   hydrometer   39
   pyknometer   39
   resonant devices   35
   sampling vessels   39
   sinker   39
   tank weight   48
   'U' tube weighing machine   39
   vibration
      gases   40, 41
      liquids   40, 41
Depth   *see* Level measurement
Doppler flowmeter   18
Doppler shift   128
Drag coefficient   143
Droplets   142, 144, 167   *see also* Particles in the flow, Bubbles

Efficiency of the flow process   161
Electromagnetic flowmeter   13, 16, 17
Enthalpy   65, 70, 149
Entropy   69
Equations:
   Beattie–Bridgman   157
   Bernoulli's   29, 53, 71
   compressible flow   98–100
   continuity   62, 197, 201, 202, 206, 211
   energy   62, 63
   ideal gas   22, 66
   laminar boundary layer   197, 201
   momentum   201, 202
   motion   197
   perfect gas   66
   steady flow energy   65–67
   state   149, 154
   turbulent boundary layer   207
   van der Waal's   155

Fanno line   114
First law of thermodynamics   63, 64
Flow   *see also* Compressible flow
   downstream of a 'T' junction   14, 166
   from an oil well   135
   in an annular combustor dump diffuser   175
   in a slurry handling pump   170
   in a small gap   91, 205
   in convergent and divergent ducts   88, 89
   in open channels   131
   low head   139
   multi-phase   *see* Multi-phase flow
   shooting   132
   subcritical   132
   subundal   132
   supercritical   133
   superundal   132
   through poppet valves   170
   two-phase   *see* Multi-phase flow
   unsteady   123
Flow-induced oscillations   127
Flow conditioners   78, 79   *see also* Flow straighteners
   K-lab   79, 80
   Mitsubishi   79, 80
   perforated plate   79, 80
   Zanker   79, 80
Flow pattern maps   137, 140
Flowmeters   *see* Bulk flowmeter, Local measurement of flow, Open channel flowmeters

Index

Flow profiles in pipes:
  distorted 58
  fully developed 58
  laminar 56–58, 60, 61
  parabolic 58, 60
  turbulent 56–58, 60, 61, 166, 208
Flow effects on:
  fan 58
  flowmeter 58
  pump efficiency 58, 142
  valves 58
Flow similarity 57
Flow straighteners 79, 80 *see also* Flow conditioners
  etoile (star) 79, 80
  tube bundle 79, 80
Flow visualization 3
  aluminium particles 3
  dye 3
  erosion-prone areas 170
  fluorescence 3
  holographic techniques 3
  hydrogen bubbles 3, 5
  interferometry 3
  schlieren 3
  smoke 3
  surface paint 3, 5
  tufts 3, 5
Fluid oscillations 121
Fluid parameters 19
  concentration 19
  density 19, 37
  level 19, 48
  pressure 19
  surface tension 19, 46
  temperature 19
  viscosity 19, 43
Fluidic flowmeter 89, 121, 125
Fluidic oscillator 125
Flumes 132, 133
Friction coefficient:
  Darcy as used by Miller 75, 116
  as used with compressible flow gas tables 75, 76, 116, 117, 181–190
Froude number 122, 132

Gas–oil ratio 139
Gas:
  entrapment 141
  ideal 22, 66
  in solution 135, 138, 139
  perfect 66

Head 28 *see also* Pressure
Head loss 73
Head loss coefficient *see also* Loss coefficient
Hot wire anemometer 6, 10, 11
Hydraulic jump 131, 133, 134
Hydrostatics 53
  centre of buoyancy 55
  righting moment 55
  stability 54
  upthrust 54, 55

Ideal gas 22, 38, 66
Impedance *see* Acoustic impedance
Incompressible fluid approximation 71
Internal energy 149
Irreversibility 100
Isentropic:
  flow 100
  relationship 211
Isobars 150
Isotherms 150
Isotropic 28
Isotropy 88

Laser doppler anemometer 6, 12
LDA *see* Laser doppler anemometer
Level measurement 48
  electrical – capacitance or resistance 48, 51
  float 48, 49
  microwave 48, 49, 51

Level measurement (*continued*)
  pressure 48, 49
  resistance 49, 51
  tank weight 48
  ultrasonic 48, 49, 51, 125
Liquid surface position sensing 51
  electrical 48, 50
  gamma ray 48, 50
  optical 48, 50
  resistance 50, 51
  thermal 48, 50
  tilt switches 48, 50, 51
  ultrasonic 48, 50, 51, 125
Local velocity measurement 6
  averaging pitot 142
  hot wire anemometer 6, 10, 11
  laser doppler anemometer
    (LDA) 6, 12
  Pitot comb 9
  Pitot tube 6, 8
  Pitot-static tube 7, 8
    National Physical Laboratory
      (NPL) design 9
    round-nosed design 9
Loss coefficient 73
  for bends 75, 76
  for inlets 75, 76
  for straight pipes 74
  for valves 77
Loss in system 77

Mach number 99, 213
Manometer 27, 31–33
Mass flowmeter *see* Bulk flowmeter
Metric standard reference
    conditions 28
Mixing length theory 208
  fan jet 208
  plane jet 208
  plane mixing layer 208
  round jet 208
  turbulent pipe flow 208
Mollier chart 153

Momentum flowmeter *see* Bulk
    flowmeter
Multi-component flow *see* Multi-
    phase flow
Multi-phase flow 135
  bubbly 136, 140
  dispersed 140, 141
  flow pattern maps 137
  from an oil well 135
  froth 141
  horizontal two-phase 136, 137
  humidity 139
  particulate matter 139
  plugs 137, 140, 141
  slugs 136, 137, 140, 141
  stratified 137, 140
  three-phase vertical 136
  two-phase 147

Non-slip condition 57, 201
Non-Newtonian fluids 45
  shear thickening 45
  shear thinning 45
Normal temperature and pressure 28

Orifice plate flowmeter 17, 142
Open channel flow 131
  hydraulic jump 131, 133, 134
  shooting 132
  subcritical 132
  subundal 132
  supercritical 133
  superundal 132
  tranquil 132
Open channel flowmeters 131
  flume 132, 133
  weir 132, 133

Parameters *see* Fluid parameters
Particles in the flow 139, 142 *see*
    also Bubbles, Droplets
  Stokes Law 142, 143
  terminal velocity 143, 144
  trajectories 170

Index

Perfect gas equation 66, 211
Pipe flow profiles *see* Flow profiles in pipes
Pitot tube 6, 8
  averaging 142
  comb 9
  Pitot-static 7, 8
    National Physical Laboratory (NPL) design 9
    round-nosed design 9
Poiseuille flow 59
Positive displacement flowmeter 13, 15, 17
Pressure 19, 27, 28, 35, 48, 70, 149
  dynamic 6
  gauge 29
  stagnation 6–8
  static 7, 8, 30
  total 6
Pressure loss coefficient *see* Loss coefficient
Pressure measurement 31
  Bourdon tubes 33, 34
  capacitance devices 35, 36
  dead-weight tester 35, 37
  diaphragm devices 33, 34
  mechanical devices 27, 33
  electro-mechanical pressure transducer 35
  manometer 27, 31–33
  Piezo electric 35, 36
  Piezo resistive 35, 36
  resonant devices 35, 36
Profiles in pipes *see* Flow profiles in pipes
Property 149
  extensive 149
Pulsation in pipes 121

Recirculation region 180
Reynolds number 53, 55
Righting moment 55
Roughness of pipes 57

Second law of thermodynamics 68
Shear layer 90, 126
Shear thickening 45
Shear thinning 45
Shock waves 100, 106–112
  equation 108, 214
  structure 107
Sound speed 97, 163, 211
  for a perfect gas 213
  in liquid 124
Specific volume 37, 38, 149
Steam 139
  critical isotherm 151
  critical point 151
  dryness fraction 140, 151
  equation of state 154
  saturated liquid 148, 150
  saturated steam 163
  saturated vapour 148, 150
  saturated vapour pressure of water 23
  saturation line 150, 151
  superheated 147
  triple point 151
Stokes Law 142, 143
Strouhal number 126
Superficial velocity 140
Surface *see* Liquid surface position sensing
Surface tension 19, 46
Surface waves 121
Surge tank 124

Temperature 19, 149
  scales 19, 20
Temperature measurement 20
  bimetallic strip 20, 21
  Bourdon tubes 22, 23
  expansion of solids 19, 20
  gas thermometer 19, 22
  liquid in glass thermometer 19, 21
  liquid in metal thermometer 19
  other liquids in glass 21, 22

Temperature measurement (*continued*)
  platinum resistance
    thermometer 19, 23, 25
  thermistor 19, 24
  thermocouple 19, 26, 27
  vapour pressure 23
Thermal expansion coefficient 37
Thermal flowmeter 18
Thermodynamic concepts:
  adiabatic 68
  control volume 65
  energy 211
  energy equation 62
  enthalpy 64, 65, 149
  entropy 69, 70
  equation of state 149
  extensive property 149
  first law 63, 64
  heat flow 64, 65
  internal energy 149
  property 149
  pure substance 149
  reversible 69
  second law 68, 69
  state 149
  system 64
  thermal equilibrium 63, 64
  work done 64, 65
Thermometer ranges 21
Tribology 92
Turbine flowmeter 17, 141, 177
Two-phase flow *see* Multi-phase flow

Ultrasonic flowmeter:
  correlation 128
  doppler 18
  transit time 18, 141
Ultrasound 121, 127, 129
  distance and position sensing 127
  doppler effect 128
  impedance 128, 129
  velocity measurement 127, 128
Upthrust 54, 55

Van der Waal's equation 155
Variable area flowmeter 17, 175
Venturi meter 17, 72
Viscosity 19, 44
  of air 44
  dynamic (absolute) 43, 44
  eddy 207
  kinematic 43
  of steam 44
  of water 44
Viscosity measurement:
  Redwood 46
  Saybolt seconds 46
Volumetric flowmeter *see* Bulk
    flowmeter
Vortex shedding 119, 121, 125, 175
Vorticity/Vortices 17, 90, 121, 125, 141
Vortex flowmeter 13, 16, 17

Water hammer 121, 123
Wave:
  length 122
  velocity 121, 123
Weirs 131–133